INVITATION TO GEOLOGY
THE EARTH THROUGH TIME AND SPACE

The Natural History Press, publisher for The American Museum of Natural History, is a division of Doubleday & Company, Inc. Directed by a joint editorial board made up of members of the staff of both the Museum and Doubleday, the Natural History Press publishes books in all branches of the life and earth sciences, including anthropology and astronomy. The Natural History Press has its editorial offices at Doubleday & Company, Inc., 277 Park Avenue, New York, New York 10017, and its business offices at 501 Franklin Avenue, Garden City, New York 11530.

William H. Matthews III is Professor of Geology at Lamar State College of Technology in Beaumont, Texas. A well-known geologist, educator, and author, he has written numerous books and articles but is best known for his very successful books, *Fossils: An Introduction to Prehistoric Life* and *A Guide to the National Parks: Their Landscape and Geology.*

Professor Matthews received his education at Texas Christian University and The University of Texas and has experience in industrial geology, research, and teaching.

A member of leading professional societies in his field —he is listed in *American Men of Science* and *Who's Who in American Education*—Professor Matthews is active in educational work, particularly in matters of curriculum planning in the field of geology and earth science. He is also Visiting Geoscientist for the American Geological Institute, Washington, D.C., and Visiting Scientist for the Texas Academy of Science.

In 1965 Professor Matthews received the Neil Miner Award of the National Association of Geology Teachers for his activity in furthering public interest in geology, and in 1966 was cited for "outstanding scholarly and academic achievement." He is a Past President of the National Association of Geology Teachers and a Fellow in the Geological Society of America.

INVITATION

TO GEOLOGY
THE EARTH THROUGH TIME AND SPACE

WILLIAM H. MATTHEWS III

PUBLISHED FOR THE AMERICAN MUSEUM OF NATURAL HISTORY
BY THE NATURAL HISTORY PRESS, GARDEN CITY, NEW YORK

The illustrations for this book
were prepared by the Graphic Arts Division
of The American Museum of Natural History

Library of Congress Catalog Card Number 70–123701

Copyright © 1971 by William H. Matthews III

All Rights Reserved
Printed in the United States of America

To Rose and Maurice J. Meyer

CONTENTS

PREFACE

Man has relied on the earth for survival since prehistoric time, and even now our destiny is inextricably linked with that of planet Earth. Yet despite our familiarity with and dependence on the earth, most of us have limited knowledge—and appreciation—of our planetary habitat.

This is an invitation to learn more about the remarkable planet that man calls home. Where did it come from? Of what is it made? What changes has it undergone during the 4½ billion years that it has been hurtling through space? Perhaps more important, what is its future—and what can we do to use more wisely the earth materials that make life possible?

But this is not a "crash" course or "survey" of geology, for this book was written to be *read* rather than studied. It is, instead, a "broad-brush" treatment of the more fundamental aspects of the study of the earth. It will, nevertheless, provide the reader with the more basic essentials of physical and historical geology and show the relation of Earth to the other astronomical bodies in the universe. It is not only an invitation to geology, but to the world of the geologist: what he is, what he does, and how he goes about doing it.

Why accept an *Invitation to Geology?* An understanding of this ancient, but relatively fragile planet, will sharpen one's appreciation for the intricate, multifaceted natural environment that Earth has provided for us. Anyone will be richer for such geological insight, for even a passing acquaintance with geology can greatly enhance one's appreciation for the exciting planet called Earth.

WILLIAM H. MATTHEWS III
Beaumont, Texas

December 10, 1969

1

AN INVITATION FROM THE EARTH

More than two million years ago in what is now East Africa, a primitive manlike creature stooped down and picked up a stone. Although it was probably only a split pebble with a sharp edge, this prehistoric man may have used his crude pebble tool to splinter a bone or to crack open a nut. Or perhaps he cut meat or scraped an animal hide with it. It is also likely that early man sought shelter and warmth in protective caves formed in the bedrock of the earth's crust. Needless to say, we cannot be certain as to the exact use he made of his pebble tool or precisely when he decided to live in caves. Nor is this important. The significant point is that he *used* them. In so doing, man had—albeit unknowingly—accepted his first invitation from the earth.

Ancient man's acceptance of this "invitation," and his use of crude tools, was an important milestone in human development. First, it clearly set him apart from the lower animals from which he came. Equally important is the fact that it indicated that he had evolved to the point where he recognized the value of earth resources—the rocks and minerals of which our planet is composed.

As time passed and man developed, he became increasingly aware of and dependent on the earth materials around him. He noticed, moreover, that the rocks beneath his feet were useful in a variety of ways. There were flat, water-worn stones in many stream beds; these made fine hearthstones for the fires that he had so recently learned to use. And flint and a type of volcanic glass called obsidian were very hard and would break with a shell-like pattern. This led to the discovery that these stones could be chipped and sharpened to make effective weapons and implements.

Clay, on the other hand, was soft and plastic, and early man learned that this common rock could be shaped and hardened to make pottery. At first clay was used to make crude vessels to store and transport food and water or to cook in. But as man's culture progressed, mere utility was not enough—he wanted to beautify his handiwork. The earth again supplied his needs, for he discovered ochers or mineral paints that could be used to decorate his earthenware. Interestingly enough, similar mineral pigments were used by cavemen as early as 16,000 B.C. They provided the color for the famous paintings of now-extinct animals that have been found on the walls of caves in southern France and northern Spain.

Some time later, ancient man discovered still more of Earth's natural treasures, for there is evidence of rather widespread use of such materials as flint, obsidian, and salt. Trading also flourished in semiprecious stones such as jadite, amethyst, and turquoise.

But it was not until about 15,000 B.C., that the first metals were recognized. These, like other mineral resources used by early man, were probably discovered by accident, presumably as native (or pure) metals deposited in stream beds. It is not known whether gold or copper was the first metal used by man. But it was probably gold, for gold is more likely to be found in the native state. Still, two of our more important metals, gold and copper, are commonly mentioned in the early history of almost all ancient civilizations. Later cultures made use of other metals such as iron, bronze, silver, and lead.

Thus, from the dawn of the Stone Age through the Age

of Metals, man has been dependent on his mineral resources. But today—in the midst of the Atomic Age—our culture and economy rely more heavily than ever before on our natural mineral resources. Indeed, this modern Age of Nuclear Energy would not be possible without energy derived from radioactive minerals that occur in the earth's crust. Clearly, man's response to Earth's invitation has shaped the course of civilization, for our destiny is inextricably linked with that of planet Earth. More important, the increasing awareness of man's dependence on this planet gradually led him to learn more about its composition and history. This need to know eventually produced results, for out of it arose a fascinating and vital new science: *geology,* the science of the earth.

2

THE EVOLUTION OF A SCIENCE

Although man soon learned to accept Earth's invitation to partake of her riches, he was slow to realize—or acknowledge—the true significance of this unique planet. In earlier times this presented no problem, for primitive man had little basis, or reason, for understanding such a great phenomenon. His knowledge of the earth was limited to the ground beneath his feet, or at most to the rather limited area that he might explore on foot. Indeed, prior to the fifteenth century A.D., no one had ever seen more than a very small segment of the earth's surface.

Nevertheless, history indicates that man's development was accompanied by an inherent and continuing desire to know more about his planetary home. Earlier interest in the earth was probably stimulated by two characteristically human reactions: curiosity and fear. It is likely that curiosity first prompted early man to pick up the unusual pebble that he used as a tool. Much later in his development, man's natural inquisitiveness would lead him to learn more about his planet, its history and composition. And, in the final analysis, it was man's insatiable need to know that caused Columbus to set sail in 1492, Galileo to develop the telescope—and the United States to place astronauts on the moon.

Natural or Supernatural?

Then, as now, man's fear of geologic phenomena centered around such violent and unexpected events as earthquakes and volcanic eruptions. Why wouldn't he have been puzzled and frightened by these devastating crustal upheavals and mountains that spew fire? This was a perfectly natural reaction, for most people are still awed and terrified by earthquakes and volcanoes. Equally normal was his attempt to explain these natural phenomena, usually by means of unworldly, supernatural forces. Although we no longer invoke myths or legends to explain them, there are still many unsolved problems as to the detailed mechanisms of both of these dynamic geologic forces.

Some of the ancients sought answers to more subtle, but equally perplexing aspects of the earth. They wondered, for example, how this planet originated and what, if any, was its relation to the sun and moon. They also puzzled over the origin of the land. And how could one possibly explain the entombed shells of sea creatures in a desert or high on a mountaintop?

At this stage of man's intellectual development, he had no means of grappling with such complex problems. And so—as they did with most natural phenomena—the ancients fabricated myths and legends to explain away the great unknown.

What caused earthquakes? According to an early Hindu legend, the earth was supported by elephants that stood on the back of a turtle, an incarnation of the god Vishnu. The turtle rested on a cobra, the symbol of water. When any of these heavily burdened creatures moved, the earth would vibrate, thus producing the temblor. The early Greek and Roman philosophers also speculated on the causes of volcanoes and earthquakes. This is not surprising, for both Italy and Greece are located in a region that is subject to both volcanic and earthquake activity. As early

FIGURE 1 Natural phenomena were first thought to be the result of supernatural forces. According to Hindu legend the earth was supported by elephants resting on the back of a turtle that was supported by a cobra. Movement of these mythical creatures would cause the earth to shift, thus causing earthquakes.

as the fourth century B.C., Aristotle taught that earthquakes were generated by pressures caused when trapped air escaped to the earth's surface. In the first century B.C., the Roman scholar Lucretius suggested that earthquakes were caused by the roof collapse of vast subterranean caverns. Although they were long since disproved, the theories of these two ancient scholars had one important feature in common: Both attempted to explain by *natural* causes events that were at that time believed to be the work of *supernatural* forces.

Because of their violent explosions and fiery outpourings, the action of volcanoes was particularly shrouded in fantasy and superstition. Indeed, the very word "volcano" reflects man's early opinion of these "fire breathing" mountains, for it comes from the Latin words *vulcanus* or *volcanus*. The term volcano is believed to be derived from the name of an ancient island off the coast of Sicily that is now thought to be the island of Vulcano. The early Romans believed that this smoking mountain was the home of Vulcan, the Roman god of fire. Vulcan was also considered to be the blacksmith of the gods, and when Vulcan's forge was heated the smoke was carried away through Vulcano's crater. The noise and vibrations from volcanic explosions were explained with equal ease: They were caused when the mighty Vulcan pounded on his anvil.

Where had the land come from? A Polynesian legend holds that one of their gods fished the earth out of the ocean, and when his line broke the land plopped back into the sea. Fortunately, it did not sink completely out of sight; the higher parts remained above the water level and these formed the islands upon which they lived. Equally fanciful was the suggestion that a turtle had dived into the ocean and surfaced with a mass of mud on its back. This lump of mud became the land.

The origin of sea shells far from the ocean caused considerable speculation among early Greek and Roman thinkers. In the sixth century B.C., Pythagoras suggested that fossilized marine shells collected in the hills of southern Italy proved that this area had once been inundated by the sea. Although quite correct, he was soundly rebuked for putting forth such a revolutionary and heretical proposal. Herodotus, about 450 B.C., made a similar observation on the remains of shellfish found in the Libyan desert. He inferred, and rightly so, that the Mediterranean Sea had once covered this region. Herodotus, incidentally, also stated that lower Egypt had been developed from sediments deposited by the Nile. Because of its roughly triangular shape (resembling the Greek letter Δ), he named this area the Delta. Not only is this name still used in Egypt today, geologists use the term delta to describe similar river-laid deposits in all parts of the world.

A Glow in the Dark

History suggests that there were few attempts to solve Earth's mysteries during the long intellectual drought of the Middle Ages. Even so, at least one glimmer of light penetrated the scientific darkness of this period and kept scientific thought alive. It was generated early in the eleventh century by a remarkable man named Avicenna. A Persian physician and Islamic scholar, Avicenna's interest in the earth was apparently stimulated by his translation of Aristotle's works. However, his writings clearly indicate that Avicenna's understanding of early geology surpassed that of Aristotle. And, in the light of modern knowledge, his interpretation of certain geologic processes seems positively uncanny. Consider, for example, this passage from one of the works attributed to this enlightened Persian:

> Mountains may arise from two causes, either from uplifting of the ground, such as takes place in earthquakes, or from the effects of running water and wind in hollowing out valleys in soft rocks and leaving the hard rocks prominent, which has been the effective process in the case of most hills. Such changes must have taken long periods of time, and possibly the mountains are now diminishing in size. What proves that water has been the main agent in bringing about the transformation of the surface, is the occurrence in many rocks of impressions of aquatic and other animals. The yellow earth that clothes the

surface of the mountains is not of the same framework of the ground underneath it, but arises from the decay of the organic remains, mingled with the earthy materials transported by water. Perhaps these materials were originally in the sea which once overspread all the land.

Here in one succinct—if somewhat poorly edited—paragraph, Avicenna recognized basic geologic processes that were not to be scientifically accepted for many centuries to come. He recognized earthquakes as mountain builders; running water and wind as erosional agents shaping the landscape and wearing away mountains; flooding of the lands by ancient seas; the development of solid rock from soft, water-transported sediment; the formation of soil; and the recognition of fossils as the remains of ancient animals. These were remarkable deductions, but Avicenna made an even more significant discovery: He clearly grasped the importance of time as a factor in geologic processes. As we shall see later, each of the ideas expressed by Avicenna have not only been substantiated, they are among the keystones of modern geologic thought.

Thus, despite the contributions of a few early scholars, their efforts were stifled by superstition and religious oppression. Consequently, their work had little influence on modern geologic thought. In fact, unlike such venerable sciences as astronomy, mathematics, and physics, geology as we know it today has developed only during the past three hundred years. More specifically, geology is largely the product of the last two centuries. Indeed, the word "geology" was coined less than two hundred years ago.

The Flame Rekindled

Man's curiosity about Earth was rekindled early in the Renaissance. The first, and probably most significant, reawakening was prompted by Leonardo da Vinci. More famous for his notable contributions to art and engineering, geologists credit Leonardo with being one of the earliest individuals to recognize the true meaning of fossils. Like the findings of Avicenna, da Vinci's comments on fossils have decidedly modern overtones—especially when one considers the beliefs concerning fossils during his time. This remarkable Italian noted that fossil shells collected in the Apennines were quite like those of similar species living in his time. He refuted the then-accepted theory that fossils were freaks of nature created by mysterious "plastic forces" within the earth's crust. Nor, he said, were they works of Satan that had been placed in the rocks to confuse or mislead those who found them. Leonardo's investigations led him to conclude that fossils were evidence of past marine life and that they proved that the present relations between lands and seas have not always been the same. He also stated that the Noachian Flood recorded in the Scriptures could not possibly have produced these fossils nor be responsible for their occurrence in the highest mountains. This was a radical and dangerous statement at a time when most Earth phenomena were being explained by literal interpretation of the Scriptures, which included the Mosaic account of Creation and the Flood. To challenge Holy Writ was an unpardonable sin, and one for which more than one early scholar was severely punished. But like most of his inventions and engineering accomplishments, da Vinci's approach to the study of the earth was far ahead of its time, and the world was not ready to accept it. Da Vinci's geologic techniques were modern in that he attempted to explain features found in the rocks in the light of natural processes in operation. In other words, he used the present as a clue to the past. This concept, as we shall see later, is one of the basic principles of modern geology.

Somewhat later, a German named George Bauer directed his attention to the study of rocks and minerals. Writing in Latin under the name Georgius Agricola, he published six books on geology. Two of these, *De natura fossilium* (1546) and *De re metallica* (1556), are considered to have laid the foundations for mineralogy and mining geology.

A Great Step Forward

The seventeenth century marked a turning point in the advancement of science. This was the age of Galileo, Newton, and Bacon, and there was a decided upturn in all fields of scientific activity. Geology advanced during this era largely due to the efforts of one man: Nicolaus Steno, Danish physician, theologian, and professor of anatomy. Credited with having written (in 1667) the first geological treatise, Steno wrote on processes of sedimentation, the origin of rocks, the formation of mineral crystals, and the interpretation of rock strata. He was also interested in the origin of fossils. Despite Leonardo's earlier conclusions, it was generally believed during Steno's time that fossils had literally grown in the rocks in which they were found. How and why had they grown there? As in earlier times, some mysterious and unknown "force" was used to explain their occurrence. Steno disagreed, and after studying fossil-bearing rocks he raised two objections. First, he could find no evidence of fossils growing in the rocks at the present time. He said, moreover, that if the fossils actually *had* grown in the rocks it would have been necessary to displace some of the rock to provide space for them. Steno could find no evidence of such deformation.

How, then, had the fossils gotten into the rocks? In his search for an answer, Steno turned to direct observation of the geological phenomena around him. And—as geologists do today—he used these to help interpret events of the geologic past. Steno observed that particles of rock sediment similar to those found in fossiliferous rocks are today being transported by currents of water. He noted further that these particles tend to settle to the bottom of quiet bodies of water and there form layers of sediment. His conclusion: Fossils represent the remains of ancient organisms that lived in the water and were covered with sediment before the sediment hardened to form rock. Steno also recorded the presence of fossiliferous strata high above present levels of the sea. He correctly reasoned that there must have been drastic changes in the sea level during earlier times.

Steno's final contribution to geology was published in 1669, six years before he became a priest in the Roman Catholic Church. Since it bears the cumbersome title *Stenonis de Solido intra Solidum naturaliter contento dissertationes prodromus,* it is not surprising that this publication is more commonly called the *Prodromus.*

Steno originally intended his *Prodromus* to be a preliminary work to a more extensive *Dissertation* that he was writing. Unfortunately, his service to the Church precluded the completion of his *Dissertation.* Even so the *Prodromus* was an important milestone in the evolution of geologic interpretation. In it he established criteria for distinguishing between marine and fresh-water sediments and further developed his earlier proposal that most layered rocks were formed by the compaction of ancient water-laid sediment. The *Prodromus* also contains the earliest known geologic cross section (FIGURE 2). Showing how a slice of the earth's crust might look if we could lift it up to view,

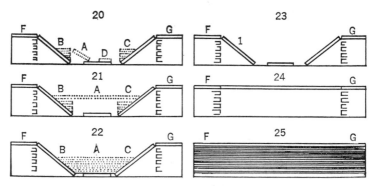

FIGURE 2   This crude cross section—one of the earliest known —appeared in Steno's *Prodromus.*

geologic cross sections are an invaluable aid in understanding the relation between exposed rocks and those in the subsurface (FIGURE 3).

More important, however, Steno's *Prodromus* clearly outlines two of the basic principles now used in interpreting earth history. He was the first to recognize that the lower layers in a sequence of rock strata must be older than those deposited above it. This is now known as the Principle of Superposition, a fundamental supposition in geology. Steno also inferred that rock strata were formed from layers of sediment that were originally laid down in a nearly horizontal position. This observation led to the Principle of Original Horizonality of strata, another basic idea in modern geology. He also correctly explained deformed and tilted strata as the result of later movements of the earth's crust. Steno's contributions to geology were truly monumental, and although they now appear to be a collection of ideas that we take for granted, in their historical setting one must marvel at the accomplishments of this perceptive physician-bishop-geologist.

A Science Is Born

Thus, from the days of the ancient classical scholars through the Renaissance, the basic tenets of geology were slowly evolving. But this progress was sporadic, consisting of scattered and often unrelated observations. Very little had been published, and though certain geologic principles had been recognized, they had not been clearly spelled out or related to each other. In short, geology lacked a basic unifying concept that might give it the status of a true science.

Then, in 1795, a book was published that provided this missing unifying thread and marked the turning point in the development of earth science. This important book, *Theory of the Earth with Proofs and Illustrations,* was written by James Hutton. Trained as a physician, Hutton lived in Edinburgh but traveled widely in northern Britain,

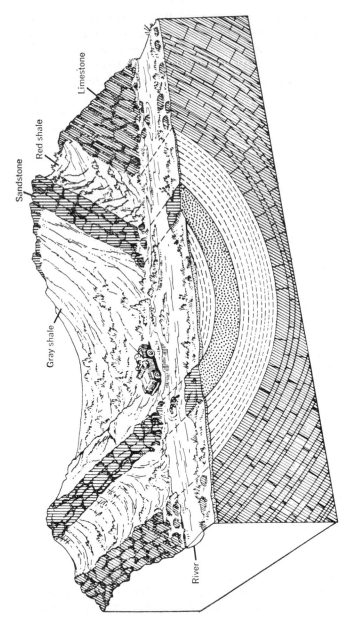

Figure 3 Geologic cross sections are invaluable aids in deciphering the geologic history of an area. (U. S. Geological Survey)

Limestone

Red shale

Sandstone

Gray shale

River

where he became interested in the earth's surface. Although Hutton's field observations were good and his interpretations remarkably accurate, acceptance of his ideas was greatly hampered by his dull and wordy style of writing. Luckily, one of Hutton's friends, John Playfair, recognized the potential of Hutton's poorly phrased contribution. Playfair, professor of mathematics and philosophy at Edinburgh, was a skillful writer with a knack for the logical presentation of ideas. His book, *Illustrations of the Huttonian Theory of the Earth,* not only clarified Hutton's statements, but included some of his own ideas. It was published in 1802 (five years after Hutton's death) and was widely read and accepted by geologists of the day.

What was the unifying concept presented by Hutton? Curiously enough, it was a principle that had been recognized and utilized by Avicenna, da Vinci, and Steno many centuries earlier. In light of evidence from his field studies, Hutton concluded that the natural processes that are now modifying the earth's surface have operated rather uniformly and continuously in the geologic past. Or, more simply stated: The present is the key to the past. Now known as the Principle of Uniformitarianism, or the Principle of Uniformity of Process, this thesis lies at the very heart of most geologic interpretation. Equally important is the fact that Hutton also clearly recognized the significance of time—immeasurably vast spans of time—in the operation of geologic processes.

With the Huttonian concept of uniformitarianism to guide them and with a better understanding of the immensity of geologic time, students of earth history were at last able to explain many of Earth's features on a logical basis. But Hutton's uniformitarian principle did more than provide the unifying concept that geology so badly needed. It is also one of geology's more outstanding contributions to modern scientific thought.

Hutton is commonly referred to as the "Father of Geology," a not undeserved title in view of his modern approach to the study of the earth. But if Hutton "fathered" geology, John Playfair was certainly the midwife who presided at its "birth." Without Playfair's enthusiasm for Hutton's ideas and his ability to express them clearly, the infant science would surely have been longer in the borning.

Following its entry into the family of science, the infant geology was carefully nurtured by a host of illustrious "foster fathers." This entire book could easily have been devoted to the important contributions of those geological pioneers, who include Buffon, Desmarest, Werner, Hauy, Cuvier, Smith, Lyell, and Darwin, to name but a few. In fact, much larger volumes *have* been written on this subject, and several of them are listed in the back of this book.

The foregoing résumé of the evolution of geology is necessarily short. Brevity notwithstanding, this historical review should serve two purposes: first, as an introduction to the development of geologic thought; second, to emphasize the relative youth of geology as a science and how far it has progressed in a relatively short length of time.

3

GEOLOGY AND GEOLOGISTS

How does one accept an invitation to the earth? By gathering pebbles like our caveman, or attempting to explain geologic phenomena, as did the ancients? The answer is "Yes" on both counts. Each year tens of thousands of "rockhounds" avidly collect tons of rocks and minerals, and geoscientists the world around are still seeking final answers to the questions that puzzled the early scholars.

But fortunately, one does not have to be a rock and mineral hobbyist or a professional earth scientist to recognize and accept an invitation to geology. In a quite literal sense geology is all around us, for the earth scientist considers geology not only as a science, but as the physical manifestation of earth materials and earth processes. Thus, the geologist speaks of the geology of a given area in terms of what is revealed in the rock formations exposed there. We say, for example, that the Rocky Mountain region is an area of complex geology. As the result of mountain-building movements, the rocks there have been deformed by folding and fracturing and severely eroded by ice, water, and wind. By contrast, the Gulf Coastal region of the United States consists largely of undisturbed, flat-lying strata, which were formed from sediments deposited within relatively recent geologic time. This is not to say that the answers to problems in Gulf Coast geology are always readily evident. They most decidedly are not, but the geology there is certainly not so structurally complicated as in the Rockies.

Despite its degree of complexity, the geólogy of any area raises many questions. What type of rocks are present, and what is their significance? Is there evidence of volcanic activity? Was the area once blanketed by great sheets of ice? Do the strata contain fossils or radioactive minerals that provide some indication as to when the rocks were formed? And are there minerals that may be of economic value? If so, what are they and how might they be recovered? The list of questions is endless, for every geological facet of an area issues its own invitation—and challenge—to those who will accept it.

Happily, an invitation to the earth is not an inducement to attend a "private party," for geology is for all who care to acknowledge it. Indeed, we could not avoid it if we wished, for in one way or another it affects every aspect of our lives. How can we accept this invitation to become more familiar with our earthly home? In a sense it is a matter of seeing rather than just looking; of understanding what we see rather than merely enjoying it. It is appreciating the value and limits of earth resources rather than taking them for granted. Not that simply looking at our surroundings should preclude enjoyment. Far from it; even a passing acquaintance with geology will sharpen one's perception and heighten his enjoyment of his natural environment.

Consider, for example, our National Parks. Most of these splendid natural preserves have been established because of their magnificent scenery and unspoiled beauty. Each year millions of visitors throng these areas, and—directly or indirectly—the scenery is the magnet that draws them there. Who, for example, does not marvel at the breath-taking vastness of the Grand Canyon, the silent mystery of a great cavern, or the thermal wonders of Yellowstone? Fortunate is the visitor who understands something of the science behind these fascinating landscapes, who recognizes the Grand Canyon as the product of an ageless contest of river against rock, Carlsbad Cavern as the imperceptibly slow work of ground water on soluble limestone, and Yellowstone's geysers and hot springs as evidence of subterranean volcanic activity.

Consider also the less spectacular but more important aspects of geology that affect each of us in myriad ways. Our very lives depend upon the products of soils that have been formed from weathered rock, and for energy we rely largely on coal and petroleum produced from the remains of prehistoric plants and animals. Think too of world industry. In addition to coal and oil, it requires ever-increasing amounts of iron, lead, copper, sulphur, salt, and other mineral products. Without these earth materials modern industry and our entire economy would soon grind to a paralyzing halt. Nor should we overlook the earth's cultural and aesthetic value. In addition to the scenery—which is too often taken for granted—Earth provides us with precious metals, beautiful gemstones, and many of the materials used in painting and sculpture. Yet these are only a minute fraction of the useful and valuable products that this planet makes available to us. These and many more earth materials have been made more readily accessible through the application of basic geology and geological engineering.

All these mineral resources and the magnificent scenery of our National Parks are the results of geologic processes that are still at work within and on the earth today. They are the same processes that began to shape our planet soon after its birth more than 4½ billion years ago.

What Is Geology?

The word geology is literally self-defining, for it is derived from the Greek *geo* (earth) plus *logos* (study). Geology, then, is the study of the earth: its origin and development through time; its composition, shape, and size; the processes that are presently or have formerly operated within it and on its surface; and the origin and evolution of its inhabitants. In their role as students of the earth, geologists make systematic observations and measurements to compile an organized body of knowledge about our globe. They study rocks, minerals, and fossils; streams, glaciers, and the sea; mountains, plains, and the moon; earthquakes, "tidal" waves, and volcanoes; and the history of life from speculations as to its origin through the appearance of man.

But the goal of the geologist is not merely to compile an encyclopedic mass of unrelated facts. Rather, he interprets these data to develop principles and hypotheses that will explain his findings and their relation to the earth. Because geology embraces such a broad spectrum of fields and scientific inquiry, geologists overlap into and rely heavily on other sciences. Basic sciences such as astronomy, chemistry, physics, and biology play significant roles in the study of the earth.

From astronomy the geologist has learned where planet Earth fits into the universe and its relation to other celestial objects. Astronomers, commonly in conjunction with other scientists, have also formulated a number of hypotheses to explain the origin of Earth and our solar system. With the advent of the Space Age, astronomy has assumed a new and more exciting role in geology. A special breed of earth scientist is now combining certain elements of astronomy and geology in the pursuit of *astrogeology*. Sometimes referred to as *planetary geology,* astrogeology requires the application of geologic techniques to the solid parts of astronomical bodies other than Earth. In practice the astrogeologist applies the principles and techniques of classical geology to space-related problems. Thus, in its broadest form, astrogeology includes the study of the form, structure, composition, history, and the processes at work on the surfaces of other solid objects in our solar system. It is unique, however, in that it differs from classical geology in a number of ways. For example, the geological processes that have sculptured the lunar landscape differ markedly from those on Earth. Consequently, surface features on the moon cannot always be interpreted in light of what we know about similar features on Earth. As the space program develops and as we look forward to landing astronauts on Mars in the 1980s, the new field of planetary geology is destined to become increasingly important.

Chemistry is of inestimable value in studying earth materials, for Earth is composed of rocks and minerals that are made of chemical elements and compounds. The geological scientist who specializes in the chemistry of rocks, their composition, and the chemical changes they undergo is called a *geochemist*. The work of these specialists straddles the boundaries of chemistry and geology. Not only are geochemists interested in the chemicals that make up the earth's crust, they are especially concerned with the distribution and migration of chemical elements in minerals and rocks. These may serve as clues to locate valuable mineral deposits. Geochemistry also plays an important role in *geochronology*—the study of time in relation to the history of the earth. Certain rocks contain radioactive elements such as uranium and thorium. Younger rocks may enclose certain radioactive isotopes (isotopes are atoms of the same atomic number, but of different atomic weight). When these radioactive materials are present, they can be used to determine the age of the rocks in which they occur.

The science of physics is used in many phases of the geological sciences. *Geophysics,* like geochemistry, is an "in-between" field that employs the techniques and concepts of both physics and geology. Geophysical investigations are largely directed toward studies of Earth's magnetism, gravity, electrical properties, and similar physical characteristics. *Seismologists*—specialists in the study of earthquakes—also rely heavily on geophysical principles and methods. *Exploration geophysics* is an important practical application of geophysics. This is geological prospecting or exploration using the instruments and techniques of physics and engineering. Various geophysical devices can be used to locate and measure differences in the magnetism, gravity, radioactivity, and seismic properties of various rock formations. These, in turn, may be clues that can lead to the discovery of valuable deposits of mineral resources. Though used in the search for both metallic and nonmetallic minerals, geophysical exploration has proved especially effective in the search for oil and natural gas.

To understand better the nature of prehistoric plants and animals, the earth scientist turns to biology. By comparing the remains of fossil species with similar organisms now living, it is often possible to reconstruct ancient plant and animal assemblages and to make logical inferences as to the conditions under which they lived. In dealing with geologic processes that shape the earth's surface and create landforms, certain principles of geography are commonly used. Geology, then, might be described as a synthesis of the concepts, laws, and techniques of the physical and natural sciences directed toward the study of the earth (FIGURE 4).

The Uniqueness of Geology

Geology has drawn heavily from the other sciences and overlaps them in many areas of geologic investigation. Yet geology is not a "synthetic" or entirely derived science, for it has certain distinct characteristics that distinguish it from other basic sciences. Perhaps the most unique of these is the concept of *geologic time*. The idea of almost unlimited time in earth history grew out of James Hutton's revolutionary Principle of Uniformitarianism, for vast stretches of time are necessary if we are to use the present as a key to the past in geology.

The Huttonian view of time was an important breakthrough in explaining major geologic features and phenomena. Prior to this, it was popular to explain these as the result of sudden, often worldwide, disasters called

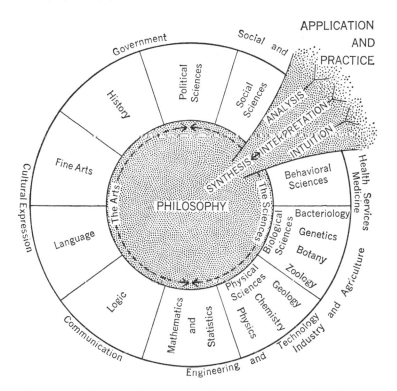

FIGURE 4 By permission from *The Sphere of the Geological Scientist* by C. J. Roy, American Geological Institute.
The interrelationships of geology with the arts and other sciences is evident in diagram.

catastrophes. According to the Doctrine of Catastrophism, valleys are openings in the earth that suddenly appeared as a result of disturbances sent down from heaven. The Noachian Flood, for example, was used to explain many geological inconsistencies that are now known to be the work of natural processes. The proponents of uniformitarianism contended that field observations failed to support the occurrence of these cataclysms. They believed instead that valleys had gradually been formed by streams that are still eroding the valley bottoms. Their conclusions were based on evidence gathered from studies of modern stream valleys in various stages of development.

Despite its importance in modern geologic thought, the Principle of Uniformitarianism should not be taken too literally. To state that geologic processes of the past were similar to those of today does not imply that they operated at the same *rate* or *scale*. For example, the rate and scale of weathering and erosion was probably much greater in very early geologic time, because Earth was not yet covered with a blanket of protective vegetation. By the same token, glacial erosion and deposition was accomplished more rapidly and extensively during the great Ice Age than it is today. In other words, although the basic nature of geologic processes has not changed, their rate and scale have varied considerably throughout geologic history. Nor should we rule out the occurrence of natural catastrophes on a local, rather than universal, basis. Surely such great disasters as the burial of Pompeii by volcanic debris and the destruction wrought by the great Lisbon earthquake of 1755 can be viewed in this light.

The significance of time in geology is clearly stated in this quote from the late Dr. Adolph Knopf, eminent American geologist:

> If I were asked as a geologist what is the single greatest contribution of the science of geology to modern civilized thought, the answer would be the realization of the immense length of time. So vast is the span of time recorded in the history of the earth that it is generally distinguished from the more modest kinds of time by being called "geologic time."*

Another distinctive feature of geology is the concept of *sequence*. Not only does the geologist want to know *when*

* Adolph Knopf, *Time and Its Mysteries*, Series III, New York University Press (1949). Reprinted by permission of the publisher.

geologic events occurred, he wants to know *in what sequence* or order they took place. The importance of sequence in geology was first mentioned by Steno in 1669. In his *Prodromus,* Steno stated that such diverse Earth features as fossil shells, mineral crystals, and certain rocks are dissimilar results of a single geologic process: the deposition of solid matter from a fluid. The concept of a sequence of geologic events assumes a rock record in which later events overlie or are superimposed on earlier events. This assumption is indispensable in interpreting earth history.

Although time and sequence are the unique themes that clearly set geology apart from its scientific neighbors, there are other differences. One of these is the problem of *scale*. Biologists, chemists, and physicists typically conduct their investigations under closely controlled laboratory conditions. But what laboratory can hold an active volcano, Alpine glacier, or Grand Canyon? Obviously these—and most other geologic features—must be investigated in the field. This literally means that the world is the geologist's laboratory, in which he must commonly conduct investigations under conditions that are often less than ideal. Fortunately, the data, photographs, and geological specimens gathered in the field can be brought back to the laboratory for additional and more specialized study.

Scope of the Geological Sciences

Because its scope is so broad and its applications so varied, geology has been divided into two major divisions: *physical geology* and *historical geology*. Physical geology is concerned with the earth's composition, its structure, and movements on and within the earth's crust. The geologic processes by which the earth's surface is, or has been, changed also falls within the purview of the physical geologist. This is a very broad division of geology, and as geologists have learned more about the physical characteristics of the earth, other and more specialized fields of

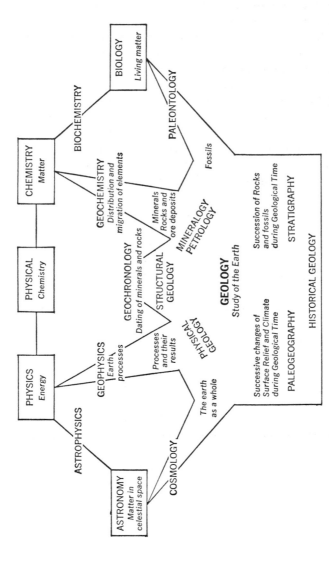

FIGURE 5 By permission from *Principles of Physical Geology* (second edition) by Arthur Holmes, The Ronald Press Company (1965). Subdivisions of the geological sciences in relation to the other sciences.

geology have developed. Thus, the *mineralogist* confines his research to minerals, while rocks are investigated by the *petrologist*. The *geomorphologist* studies the earth's surface features and tries to explain the nature, origin, and development of landforms. On the other hand, the *structural geologist* attempts to account for the various types of rock deformation that have taken place within the solid earth. *Seismologists* are interested in earthquakes and the earth's interior, *volcanologists* investigate the activities of volcanoes, and *glaciologists* study glaciers. Even more specialized are the previously mentioned *geochemists* and *geophysicists,* who use their skills to analyze the earth in the light of data derived from chemistry and physics.

The potentially disastrous combination of the continued abuse of our environment and a rapidly expanding world population has disturbed many geological scientists, and their concern has given rise to a new branch of geology. Broadly referred to as *environmental geology,* this important division of earth science is dedicated to the conservation of earth resources and the application of geology to human needs. The major goal of the environmental geologist is to make others aware of the need to consider geologic problems in planning for the future. For example, much of the damage sustained in the highly publicized California landslides could probably have been prevented had geologic advice been sought—and heeded—prior to residential construction. Thus, earth scientists can be of great help in urban planning if their services are used. The environmental geologist is also concerned with the mounting problem of waste disposal. Not only are waste products continually increasing in volume and variety, they are becoming more and more poisonous and contaminating. The disposal of radioactive products in the Denver area is thought to have triggered the unexpected earthquakes that began in 1962. Expert geologic consultation might have prevented these quakes. But though the disposal of radioactive materials will become increasingly difficult to handle, it is the wastes generated by the human inhabitants of our expanding cities that create the most serious problem.

Like physical geology, there are also specialized fields of historical geology. One of these, *cosmogony*, is related to astronomy. The *cosmogonist* makes speculations regarding the origin of the universe and the birth of our planet. The *paleontologist,* by means of fossils buried in the earth, describes and classifies prehistoric organisms and traces their rise throughout geologic time. The *stratigrapher*—a master sleuth in extracting long-buried secrets from the rocks—uses his geologic findings to place geologic events in their proper chronological order. Each of these earth historians plays an important part in deciphering the geologic history of our planet.

Whereas the physical geologist concentrates on the physical aspects of our globe, the historical geologist is concerned primarily with its history. He wants to know when and how Earth was formed, what changes it has undergone, and how life has developed on its surface. To answer these—and even more puzzling questions—the geologist must turn detective. Like the sleuth who combs the scene of a crime searching for bits of incriminating evidence, the geologist probes the rocks for clues that are equally enlightening. He is hunting for evidence that will help solve the ancient mysteries of the geologic past. Luckily, this search has been rewarding. Geologists have traced the development of the earth from its cloudy beginnings some 4½ billion years ago, through the terrible "Age of Dinosaurs" and into the frigid saga of the "great Ice Age." They have also followed the slow march of life from the earliest known simple plantlike organisms of more than three billion years ago, through an amazing array of ancient life forms, all the way to ancestral man.

The major branches of geology overlap in some areas and are interdependent. The physical geologist uses mineralogy and petrology to ascertain what types of rocks are present and from what they were derived. The historical geologist may study the same rocks to determine what kinds of plants and animals were living at the time the rocks were formed, the environment that they inhabited, and the type of climate that was present. Thus, the unification and interaction of physical and historical geology leads ultimately to a better understanding of the composition and history of the earth. This is, of course, the major objective of all geologists.

4

PLANET EARTH:

WONDER OF THE UNIVERSE

Earth—planet No. 3 in order from the sun—is a remarkable astronomical body. It is not the largest or heaviest member of our solar system; nor is it the fastest or slowest. Indeed, Earth has even been called a "second-rate" planet. Why, then, has it also been referred to as the "wonder of the universe"? Among the billions of celestial objects in the universe, only Earth is definitely known to support life as we know it today.

Earth's Companions in Space

The earth, its relation to the stars and planets, and speculations as to its origin attracted man long before the birth of the geological sciences. The Sumerians, who provide us with the earliest known accounts of the earth, saw it as a creation of the gods. To the Babylonians the earth was a hollow mountain held up and surrounded by the sea. The solid firmament, through which moved the sun, moon, and stars, was arched over it. Earth's hollow interior was the underground world of the dead. The Egyptians envisioned the earth as a recumbent, plant-covered god named

FIGURE 6 The ancients viewed the Earth in varying ways. They conceived the heavens as a bent-over goddess supported by the god of the atmosphere. The two boats were used by the sun god to cross the heavens.

Keb. The heavens were thought to be a gracefully arched goddess supported by the god of the atmosphere.

Thanks to the science of astronomy, we now have a more accurate idea about the nature of planet Earth. Much of this information has been useful to the geologist, for it is helpful to know something about the relation of Earth to other objects in the universe. The universe, as we now know it, consists of countless disk-shaped clusters of stars called galaxies. Astronomers estimate that there are many galaxies present in outer space, but only one—the Milky Way—need concern us here. This galaxy, which contains every separate star that can be seen with the naked eye, is lens-shaped and contains billions of stars including our sun. Not especially large as galaxies go, the Milky Way is nonetheless almost unconceivably large. Because of the immensity of space, astronomers measure distances in terms of light-years, the distance light travels in one year at the speed of 186,234 miles per second. We might better appreciate the distance involved if we consider that one light-year is slightly less than six thousand billion miles. How fast is that? A beam of light sent from the Pacific coast across the United States to the Atlantic (some three thousand miles) and back again would complete about thirty-one round trips in just *one second!* Consider then, the Milky Way: It measures roughly one hundred thousand light-years from edge to edge, and its thickest part is some twenty-five thousand light-years across.

Located about halfway between the center and the edge of the Milky Way is a rather ordinary star of modest size and brightness. This great mass of gaseous material is important to man, for it is the sun—the center of our solar system. The sun is not only important because it makes life possible on Earth, it also accounts for most of the continuing energy supply at the earth's surface. This is of great geological significance, for the sun's energy heats the atmosphere, the oceans, and the solid earth. The energy that it creates moves earth materials and changes the earth's surface features. We will have more to say about the sun's energy output later.

Our sun is not alone in the Milky Way. Speeding endlessly around it are the nine planets of our solar system. In order of their distance from the sun, these are Mercury, Venus, Earth, Mars, Jupiter, Saturn, Uranus, Neptune, and Pluto. The first four are called the terrestrial planets, because they are about the same size, and their density suggests a composition of iron and stone. Thus, Mercury, Venus, and Mars are believed to be similar to Earth. The four have also been referred to as the inner or minor planets. Next in order of distance from the sun are the outer or major planets: Jupiter, Saturn, Uranus, and Neptune. Pluto, which resembles the terrestrial planets, is usually treated as a separate case.

Most of the planets are associated with smaller bodies

called satellites or moons, which revolve around them. Some planets (Mercury, Venus, and Pluto) have no known satellites. Jupiter, on the other hand, has twelve. Earth's satellite, the moon, revolves around our planet once every month. Our moon has received much attention within the past several years and, as we shall see later, it may serve as the key to unlock some of the universe's age-old mysteries. Also located in our solar system are other celestial bodies, including asteroids (or planetoids), meteors, and comets.

Planet No. 3

From the above it is obvious that Earth's place in the universe is indeed minute and, compared to the sun and many of the other planets, it is not too large in size. Even so, Earth is a sizable object, for it has a polar diameter of about seventy-nine hundred miles, and its equatorial diameter is some twenty-seven miles greater. The circumference of the earth is 24,874 miles, and its surface area comprises roughly 197 million square miles. Its volume is somewhat more than 250 billion cubic miles, and its mass (or weight) has been estimated to be sixty-six hundred trillion. In shape, the earth has the form of an oblate spheroid. That is, with the exception of a slight flattening at the poles, the earth is nearly spherical or ball-shaped. This flattening and the earth's equatorial bulge are caused by the centrifugal force of rotation. How noticeable is Earth's "midriff" bulge? On a model Earth of twenty-five feet in diameter, one could hardly see it, for this slight protuberance would amount to less than one inch. On the same scale, our highest mountains would rise less than one-fourth inch above sea level.

But despite its relatively unassuming statistics, ours is a most fortunate planet that seems "just right" in many ways. Its general size, composition, daily rotation, and its distance from the sun provide optimum conditions for the development and continuation of life. The fact that Earth is the only planet that appears to have a large supply of water has undoubtedly had much to do with this. Earth's No. 3 position is also very important, for at ninety-one million miles from the sun it is not too near, nor is it too far. If, like Mercury, the earth were only thirty-six million miles from the sun, it would be subjected to temperatures as high as 640° Fahrenheit, and the seas would literally boil away. On the other hand, the temperature on Pluto, some 3 billion 670 million miles from the sun, is incredibly cold —perhaps as much as 350° below zero. Earth's average temperature of 57° Fahrenheit appears to be "just right" for life conditions.

Most of us take day and night for granted; we even have a saying "as surely as day shall follow night." But why *does* day follow night? More important, what would happen if day did *not* follow night? These questions emphasize the importance of Earth's daily rotation. Our planet rotates on its axis once each twenty-four hours, providing relatively short periods of light and darkness. This is important, for during each twelve-hour period there is not time for excessive heat to accumulate as we face the sun, or for too much heat loss as we are turned away from it. This rate of rotation, then, helps to prevent unbearable extremes in temperature. We can appreciate the ideal timing of Earth's rotation by comparing it with the moon's rate of rotation. The day is fourteen days long on the moon, and during this time the temperature rises to some 214° Fahrenheit *above* zero. Needless to say, only very especially adapted organisms could possibly exist in such an inhospitable environment. We now know—thanks to lunar material returned by Apollo astronauts—that life forms have apparently been unable to live under these extreme conditions. When we consider that both the moon and Earth are about the same distance from the sun, we can better appreciate the fact that our planet does not rotate more slowly.

Suppose, however, that our rate of rotation should suddenly become slower. Would the seas boil away and the continents melt? This would depend, of course, upon how *much* it slowed. But generally speaking, we would still not be as greatly affected as would the moon if it should slow down proportionately. This is because Earth has an atmosphere—a two-hundred-mile thick envelope of gas. The atmosphere encircles our globe, providing a protective mantle to shield us from temperature extremes such as those on the moon.

The earth turns on its axis (the shortest diameter connecting the poles), and as it rotates the earth has a single wobble. This is because the earth's axis is tilted at an angle of 23½°. Known as precession, this slight wobbling motion is so slow that it takes almost twenty-six thousand years to complete a single wobble. The inclination of the axis is also responsible for the seasons. During one-half of the year (summer and spring), each hemisphere is tilted toward the sun. During the remainder of the year (fall and winter), each hemisphere is tilted away from the sun.

In addition to spinning on its axis, the earth revolves around the sun in a slightly elliptical orbit every 365¼ days. During this time—a solar year—the earth races along at more than sixty thousand miles per hour at an average distance of ninety-three million miles from the sun. Not only does our planet move, the entire solar system is hurtling through space at a velocity of about twelve miles per second. It is heading in the general direction of Vega, the second-brightest star visible in northern latitudes.

Whence Came the Earth?

Astronomers have some indication as to the direction our planet is headed as it travels through the universe, but they are much less sure of where it came from. Where, when, and how did Earth originate? Man has speculated on these questions since the dawn of recorded history. Today, even as we probe the reaches of outer space, we continue to speculate on the origin of the earth. Thanks to the cosmogonists—scientists concerned with the origin of the universe—we now have a number of hypotheses to explain Earth's creation. But unfortunately, none of these ideas is entirely satisfactory, and serious objections have been raised against each of them.

The purpose of any theory or hypothesis is to fit together all observed facts. Thus, a workable theory of cosmogony must account for the present composition and mechanics of our solar system. On the basis of the observable facts, cosmogonists have developed two basic classes of hypotheses. Those that suggest that the planets were formed without intervention from forces outside the solar system are called single-body hypotheses. These theories assume that no object other than the sun was involved in the birth of our solar system. Second-body hypotheses assume that the solar system originated as the result of forces created by the accidental interference of another celestial object. Because the latter hypotheses generally presume that another star passed close to the sun, these have also been called two-star, intruder, or encounter theories.

One of the first scientific attempts to explain the earth's origin was made in 1755 by Immanuel Kant, a German philosopher. Later, in 1796, Pierre Laplace, a French mathematician, developed the theory more fully, and stated it in more scientific terms. Interestingly enough, Laplace was unaware of Kant's earlier work and both arrived at similar conclusions quite independently. Known as the Nebular Hypothesis, this idea assumes that the solar system originated from a large mass of gas that was spinning through space. This gaseous mass, called a nebula, gradually became smaller in response to its gravitational pull, and as it contracted its speed of rotation increased. Eventually the outermost part of the nebula rotated so rapidly that rings of gas separated from the shrinking nebular body. These rings slowly condensed to form the planets, and the central mass became the sun. The major objection to this hypothesis is that the sun is rotating too slowly in comparison with the planets for this theory to be plausible. Moreover, such rings of hot gas would probably have been dissipated in space rather than condense into solid planets.

Another theory, called the Planetesimal Hypothesis, was proposed in about 1900 by Forest R. Moulton, an astronomer, and geologist Thomas C. Chamberlin. These two University of Chicago professors theorized that the sun was a star that existed before the planets were formed. At some time in the remote past, an "intruder" star passed by the sun, thereby exerting a gravitational force strong enough to pull masses of solar material from opposite sides of the sun. This material later cooled and condensed into solid particles called planetesimals, and these small chunks of matter orbited around the sun in the same direction that the intruder had moved. The largest of these planetesimals acted as nuclei that attracted other planetesimals, and by accretion the planets eventually grew to their present size. The satellites were formed from other clumps of planetesimals located near the clusters from which the planets were formed. This second-body hypothesis explains the development of the planets, satellites, and asteroids, and overcomes the problem of a slowly rotating sun. Nevertheless, it has been discredited because astronomers believe that it is highly unlikely that another star could have passed this close to the sun. But even if this "near miss" had occurred, it is doubtful that the intruder could have generated sufficient lateral thrust to set the torn-out solar masses in orbit. Instead, they would have fallen back into the sun.

Within recent years cosmogonists have taken a closer look at a possible nebular origin of the Earth and devised a new theory in the light of newer findings. This idea has replaced the original Nebular Hypothesis of Kant and Laplace as well as the later Planetesimal Hypothesis. The revised theory, known as the Protoplanet Hypothesis, was first proposed in 1944 by C. F. Von Weizsacker and modified by Gerald P. Kuiper. It has been found that rapidly rotating nebula will develop large whirlpools or vortexes at various places on the disk of nebular material.

Each of these great whirlpools might then have collected
the surrounding material by gravitational attraction, thus
forming a protoplanet. It is believed that nine protoplanets
—one for each of the present-day planets—were formed,
and these were originally much larger than the finished
planet. Smaller whirlpools developed inside some of the
larger vortexes, and these gave rise to spinning disks that
became the satellites, or moons, of the planets. Many
astronomers support this theory because observations with
large telescopes have revealed numerous true nebulas between the stars. Equally significant is the fact that some
of these massive swirls of gas and dust are actually con-
densing to form new stars. The Protoplanet Hypothesis is
generally acceptable to most cosmogonists because it explains many of the known facts about the solar system.
It is, nonetheless, far from complete, and the origin of the
solar system and Earth is still largely in the realm of
speculation.

The "End of the World"—When and How?

This, then, is Earth. A relatively small, not-quite-round
sphere of solid, liquid, and gas that is revolving, rotating,
and speeding endlessly out into space. What is the future—
and fate—of this unique planet and its precious cargo of
life? No one knows for sure, of course, and time alone
will tell. In the meantime, man has not been content to
"wait and see." Yielding to his inherent curiosity, he has
theorized on the end just as he has speculated on the origin
of the earth. And, as might be expected, his predictions for
the end of the world are every bit as diverse as his ideas
about its beginning. At one end of the spectrum are those
prophets of doom who make all manner of dire predictions
about the end of the world and all mankind. One year they
will forecast universal floods or fires; the next a worldwide epidemic of some dread disease; more recently, earthquakes and nuclear blasts have been popular villains.

But while the Cassandras have gazed into their tea cups

and crystal balls, the astronomers have not been idle. They have looked into their telescopes and probed the universe for a more feasible prediction of our planet's grand finale. What have they learned? Though they do not pretend to tell us exactly when and how the end will come, astronomers have made some scientific prognostications on the basis of studies of older stars that resemble our sun. They have seen what happens when aging stars begin to run out of the nuclear fuel that makes them burn, and they assume that our sun will react in a similar fashion at some point in the distant future. Earth's destiny is inexorably linked with the fate of the sun, for we rely on light and radiation from our parent star. Thus, changes that affect the sun must eventually affect the earth. Indeed, if our planet was not so dependent upon the sun, the earth might possibly last forever.

When will the earth feel the impact of these predicted solar changes? The world's present inhabitants have little cause for worry, for it seems possible that life will thrive on Earth for at least five billion years—perhaps even longer. This life-span is based on the assumption that medium-size stars like the sun radiate at least ten billion years before the hydrogen fuel content of their core is exhausted. Thus, the sun, which is about five billion years old, is in "middle age" and in midstream of stellar evolution.

How will the end come? Some astronomers think that once the sun's original hydrogen has been converted into helium, its core will begin to contract. As the core shrinks, it will generate sufficient heat to trigger thermonuclear reactions capable of producing incredibly high temperatures. These nuclear furnaces should cause the body of the sun to expand, and the amount of heat and light radiation will gradually increase. Accelerated radiation will cause the sun to become redder and redder until it becomes a red giant— a mass of red-hot, rarefied gas. As the sun expands, the solar atmosphere will literally swallow Mercury and Venus as these two inner planets are vaporized by the red giant's great outpourings of heat. Although Mars and Earth could meet the same fate, their greater distance from the sun may prevent their being converted into planetary vapors. Even so, Earth will not escape the death-dealing solar temperatures—the oceans will boil and the rocks of the lands will become red-hot and some may melt away.

The evolution of our sun up to this point is fairly predictable, but astronomers are not so sure about the latter phases in the death of a star. It seems likely, however, that after its climax as a red giant, the sun will steadily diminish in brightness and energy and end its days as a white dwarf. These are stars that have completely exhausted their supply of hydrogen and other nuclear fuels and shine only because of their internal heat. As a white dwarf the sun will continue to radiate its waning supply of energy and will gradually cool and contract. In due time it will probably shrink to the size of a planet and become cold and dark. With its fires gone out, the sun and other members of the solar system will be at the mercy of the icy temperatures of interstellar space. At hundreds of degrees below freezing, gases will be converted into liquids, liquids will become solids, and Earth will eventually become a dark, frozen mass. But even in death the sun will not relinquish its hold on its companions in space. The cold, dead, inert planets will continue to pursue their never-ending travels around the cold, inert sun, because its mass and gravitational force will remain strong enough to keep the remaining planets locked in eternal orbit around it.

5

PORTRAIT OF A PLANET

As satellites and spacecraft have probed ever-deeper into outer space, we have gained a completely new perspective of our planetary home. People all over the world have thrilled to the spectacular television pictures beamed earthward by Apollo missions. And yet as beautiful and exciting as they are and despite our pride in such great technological achievements, these remarkable images have produced a sobering—even humbling—effect on man. They have emphasized as never before what a tiny part of the universe Earth occupies and how small and restricted is the environment available to us.

Earth scientists have been particularly interested in these revealing photos, for they have provided a broad field of view never before possible. For the first time, students of the earth have seen more clearly certain geographic and geologic features, and they now understand better their relation to each other. Yet most of what we know about the earth has been learned by geoscientists working on the earth's surface in close contact with rocks, minerals, and fossils. It was natural for early workers to concentrate on the composition of the earth, for any clear understanding of our planet must necessarily start with some knowledge of the materials that it is made of.

Air, Water, Land, and Life

What is the composition of the earth? The average person would probably answer "rocks and soil," for most of us think in terms of so-called terra firma: the solid earth. But solids are not the only types of matter present on Earth. What of the gaseous and liquid matter—the air and water so essential to life and geologic processes? The earth scientist is also vitally interested in these, for they are as much a part of Earth as the solid surface upon which we live.

These three types of matter—gases, liquids, and solids—are generally described in terms of three rather distinct zones or spheres. There is the *atmosphere,* a thick, life-giving envelope of gas that completely surrounds our globe; the *hydrosphere,* the far-reaching sheet of water that fills Earth's larger depressions, and blankets almost 71 percent of the earth's surface; and the solid "rocky" part of the earth called the *lithosphere.* Although these three zones of matter are clearly defined in composition, the boundaries between them are not as distinct. Rather, there is a ceaseless interaction between them as air comes in contact with rock, rock with water, and water with air. It is this continued intermingling of earth material that attracts the geologist, for important geologic changes commonly occur at the interface or boundary between these zones.

The gaseous part of Earth extends for hundreds of miles above sea level. Composed largely of nitrogen (78 percent) and oxygen (almost 21 percent), the atmosphere is a mixture of water vapor (derived from the hydrosphere) and tiny dust-size fragments of the lithosphere. Exceedingly small amounts of argon, neon, helium, and certain other rare gases are also present. This thick mantle of gas provides the air that we breathe and acts as an insulating blanket to protect us from the sun's intense heat and dangerous radiation. Also important is its function as a protective shield to fend off meteors that might otherwise bombard the earth's surface. When meteors encounter the atmosphere, friction-produced heat causes them to burst into flame. Luckily, most of these fiery "falling stars" are consumed before they reach the surface of our planet. Wherever air touches water, the atmosphere is in continual interplay with the hydrosphere. Thus, the atmosphere has a significant role in the hydrologic, or water cycle, for it is responsible for the evaporation and precipitation of moisture and the transfer of water to the land. The atmosphere also interacts with the lithosphere to produce weathering. This important geologic process is forever at work on the earth's surface, and weathering has played an important part in producing much of our present-day landscape.

Water, like air, is necessary for the existence of life. Most of earth's water is found in the extensive universal sea that covers 70.8 percent of the earth's surface to an average depth of some 2½ miles. However, the hydrosphere does not consist solely of salt water. It also includes the fresh water found in streams and lakes as well as that in the ground. The "water sphere," like the atmosphere, constantly interacts with earth's rocky crust. The sea perpetually wears away the land, and deposits sediment on the ocean floor; streams steadily widen and deepen their channels by erosion, and little by little, groundwater dissolves minerals from the buried rock formations through which it flows. In short, water—with a valuable assist from weathering—has been the major force that has shaped the face of the earth throughout geologic time.

Thanks to the life-giving qualities of air and water, our planet is populated by myriad plants and animals. This great swarm of organisms makes up the *biosphere,* an organic realm of the earth that is just as important as Earth's physical or inorganic zones. The sphere of life is intricately interrelated with the air, land, and water, for most organisms live in the narrow zone where these gases, solids, and liquids meet. There are few regions on Earth that do not support some type of life. Life forms have been found in the upper reaches of the atmosphere and on the floor of the deep ocean trenches miles below sea level. Some organisms have become adapted to the temperatures of the polar regions; others thrive in the water of hot springs or on burning desert sands. And the oceans literally teem with countless species of plants and animals, while much of the land is covered by vegetation, which supports many forms of animal life. Virtually all of these organisms are confined to the rather limited habitat provided by the narrow interface between land and water, water and air, and air and land.

The biosphere admittedly falls more properly within the scope of the biologist or life scientist. But the long-standing and continual interaction between the earth's organic and inorganic spheres make life no less important to the earth scientist. During the more than three billion years that life has been present on earth, it has spread to all parts of the world. And throughout this portion of geologic time the plants and animals of the biosphere have interacted with the atmosphere, hydrosphere, and lithosphere to become involved in a number of earth processes. The biosphere is responsible for most of the oxygen we breathe, the formation of petroleum and coal, and many rocks of organic origin. Moreover, the record of past life on earth, as revealed by fossils, has provided valuable clues to the history of the earth.

Despite his dependence on and his interest in earth's air and water, and the fact that he is part of the biosphere, the geologist is primarily concerned with the lithosphere. Here are the minerals that make up the rocks that comprise the continents and ocean basins. Here, too, are the soils, metals, and other mineral resources so vital to man.

Rocks and Minerals: Building Blocks of the Earth

Rocks, like air, water, and many other familiar objects, are too often taken for granted. We step on boulders to get across a stream, there may be gravel in our driveway,

or crushed stone on the roof of our house. Bricks and tile
are made of clay, glass from tiny grains of sand, and the
life-giving soil is a product of weathered rock. In short,
rock is everywhere around us, for it is one of the most common things on earth.

But what is a rock? The word "rock" means different
things to different people. The quarryman sees rock as a
product to be excavated and sold. To the civil engineer it
provides a firm foundation upon which to erect buildings
and build roads, while the construction worker considers
rock something to be attacked with pick and shovel during
the course of a hard day's work. The geologist, on the other
hand, looks at rock from *all* of these viewpoints, plus many
more. Generally speaking, rocks are the major units that he
studies in the field and laboratory. His investigations of
rocks permits him to distinguish one geologic formation
from another and to determine their distribution in a given
area. He uses this information to prepare a geologic map
that will help him to reconstruct the geologic history of
the area and that may reveal the presence of valuable
mineral deposits. But in the final analysis, the geologist sees
rock in its most fundamental and basic form: the stuff of
which the solid earth is made.

If the solid earth consists of rocks, what are the rocks
made of? Regardless of shape, color, size, weight, or hard-
ness, all rocks have one thing in common: They are composed of minerals. Rocks are naturally formed aggregates
of minerals; and a mineral is a naturally occurring inorganic substance that has a fairly definite chemical composition, distinctive physical properties, and that typically
occurs in definite shapes called crystals. Mineralogists have
recognized and described more than two thousand different kinds of minerals. These vary greatly in their
physical characteristics, and these differences help to distinguish one mineral from another. Minerals also differ
chemically. Some of them—native gold, for example—
consist of but a single chemical element. Others, such as
halite, or rock salt, contain two or more elements combined

to form a compound. Although chemists have identified over one hundred elements, more than 98 percent of all minerals are composed of only eight elements: oxygen, silicon, aluminum, iron, calcium, sodium, potassium, and magnesium. Equally interesting is the fact that more than three-fourths of the earth's crust consists of but *two* elements: oxygen and silicon.

Both metallic and nonmetallic minerals occur in Earth's rocky crust, and both types are equally important, for they constitute the raw material from which most rocks are formed.

The Kinds of Rock

Rock, like minerals, differs greatly, and each has its own distinctive characteristics. Rock also differs in the way it occurs in the earth's crust. It may be found as loose, unconsolidated surficial material such as soil, sand, or gravel. This is the mantle rock or regolith, a term that literally means "blanket rock." Underlying this "rock blanket" is the bedrock—a continuous mass of solid rock that has not yet been disturbed by surface agents such as weathering or erosion. Despite their difference in appearance and mode of occurrence, the bedrock and regolith are very closely related. One is actually a product of the other, for the regolith is composed of rock debris derived from the weathering and erosion of once-solid bedrock. By the same token, the loose rock fragments of the regolith may eventually become tightly consolidated and be changed back into bedrock. The cementation of loose sand grains to form sandstone is an example of the latter.

The major processes that form bedrock operate in two particular geologic environments. One of these is located within the earth; the other is found at or near Earth's surface. Thus, although there are many varieties of bedrock, virtually all of them will have originated in one of these rock-forming environments. This has led geologists to a genetic classification consisting of three major rock groups: igneous, sedimentary, or metamorphic. Although geologists may differ on some points, they generally agree that most rocks can be fitted into one of these broad classes of rocks.

But this has not always been so, and herein lies another interesting chapter in the history of geology. It dates back to the latter part of the eighteenth century, a time when two rival schools of geologists sought to explain the origin of the earth's rocks. On one side were the Neptunists led by Abraham Gottlob Werner, a pioneer German mineralogist who taught at the Mining Academy of Freiberg. The Neptunists—a name derived from Neptune, Roman god of the sea—held that all bedrock had formed from water. These rocks were assumed to have been precipitated from a universal sea that once covered the entire earth. As proof of this universal flood, they pointed to the fossilized remains of marine plants and animals in many of the rocks. Nonfossiliferous rocks such as granite and basalt (now known to have formed from a molten state) were explained away as the oldest deposits in the original ocean. The Neptunists were also quick to point out that these rocks are commonly sandwiched between other rocks that contain marine fossils, so they must have formed in the sea also. The basic tenets of Neptunism were developed in a book by Werner published in 1787, and it is not surprising that his ideas were popular in that day, for it seemed to have supported Scriptural interpretation, and especially the Noachian Deluge. The widespread waters of Noah's Flood were thought to have provided the great muddy ocean from which the rock had precipitated. Better yet, it helped to clear up the troublesome question of fossils. These stony remains were considered to be victims of the Flood, whose shells and bones were left buried in the mud as the flood waters receded.

The opposing school of thought consisted of geologists who believed that a great deal of bedrock, especially granite and basalt, had been produced by volcanic activity. Appropriately named after Pluto, Greek god of the lower world and keeper of Earth's internal fires, the Plutonists asserted that much of the bedrock had solidified from molten rock material (called magma) that rose from the depths of the earth. Because much of this rock was associated with volcanic activity, the Plutonists have also been called "Vulcanists." James Hutton, leader of the Plutonists, based his beliefs on field observations made during his far-ranging travels. But Werner had limited financial resources, and could ill afford lengthy field excursions. Consequently, his conclusions were based on rather localized geologic phenomena. Meanwhile, other European geologists had also taken to the field. As early as 1751, the pioneer French geologist Jean Guettard recognized the evidence of ancient volcanic activity in the Auvergne region of France. A year later, convinced of the volcanic origin of the Auvergne landscape, Guettard published a report stating the then unheard-of opinion that there had been active volcanoes in the heart of France. More important, he was able to prove that some of the rocks in the ancient lava flows were composed of basalt. Although he apparently failed to recognize the basalt as the product of ancient volcanic activity, his conclusions were, nevertheless, quite remarkable, for Guettard had never before seen a volcano until he went to Auvergne. But this pioneer field geologist had done his "homework" well, for he correctly interpreted what he saw in the field.

Despite the work of Guettard, it remained for Nicolas Desmarest to prove the volcanic origin of the Auvergne basalts. In a field study undertaken some ten years after Guettard's pioneering efforts, Desmarest mapped a rather large area and proved conclusively that the lava flows he discovered had been produced by volcanic activity. What about lavas and basalts found in areas where volcanoes no longer existed? Desmarest correctly concluded that the volcanic cones had later been removed by erosion. But ironically, it was Leopold von Buch—one of Werner's most brilliant students—who helped administer the *coup de grâce* to Neptunism. One of the earliest authorities on volcanism, von Buch studied many volcanic regions, both active and extinct. The results of his widespread and exhaustive investigations firmly convinced him of the volcanic origin of basalt and tipped the scales in final favor of the Plutonists.

As time passed and more geologic data was gathered, Werner's faithful band of Neptunists gradually dwindled away as it became increasingly obvious that all rocks could not possibly have had an aqueous origin. Be that as it may, we should not minimize the contribution of one of the most dedicated pioneers in the field of geology, and more especially, mineralogy. As Kirtley Mather so aptly wrote of Werner in his *Source Book in Geology:* "His Neptunist school may have retarded geological thought, but his inspiring teaching and his earnest effort to classify all data did even more to advance it."

And so today we divide the earth's rocks into three main groups according to the way in which they have formed: igneous, sedimentary, and metamorphic. Rocks produced by the cooling and solidification of magma (molten rock) from within the earth's crust are called *igneous rocks.* The ancestors of all other rocks, they get their name from the Latin word *ignis,* meaning fire. These "fire-formed" rocks become more abundant deep in the earth, for some 95 percent of the outermost ten miles of the earth's crust is composed of such rock. Some igneous rocks such as granite and gabbro originated in the interior, and have formed from magma injected into the surrounding rocks. These are called intrusive, or plutonic, igneous rocks. Because they cooled far beneath the earth's surface, these rocks are typically seen in areas that have been greatly eroded. With the passage of time, the massive buried plutons have been gradually exposed as the overlying strata have been stripped away by wind, water, and weather. You will see rock of this type exposed on the bald summit of Pike's Peak, in the ice-carved spines of the Tetons, the inner gorge of the Grand Canyon, and along the rock-ribbed coast of Maine. Yet despite its widespread occurrence in many parts of the world, the origin of granite is still something of a mystery.

It is generally regarded as an igneous rock, but some petrologists are not so sure that this is correct. They argue that certain masses of granite rocks are much too large to have been injected into the surrounding rock. They cite, for example, the Coast Ranges of British Columbia: a vast linear belt some one thousand miles long and about 150 miles wide that is known to be seven thousand feet thick —and with no indication of any bottom to it. An igneous intrusion of this magnitude seems unlikely, and this has led some geologists to believe that these igneous-looking rocks were formed by the process of granitization. Unfortunately, this process is not thoroughly understood—if indeed it takes place at all—and the final solution to the origin of granite lies tightly locked within the mineral grains of this commonest of all "igneous" rock.

Desmarest proved that molten rock can also spill out on the earth's surface to form extrusive igneous rock or lava. These volcanic rocks may force their way upward to erupt from volcanic craters or flow out through massive fissures in the earth's surface. Volcanic eruptions through a central vent in the crust normally build volcanic mountains or cones. Such magnificent peaks as Mount Rainier, Mount Shasta, Mount Fuji, and a host of others have formed in this way. However, if the lava is extruded from a fissure or crack in the ground or mountain side, it spreads out rather evenly over the surface in thick, horizontal layers. Striking evidence of such *fissure flows* can be seen in the Columbia Plateau of Washington, Oregon, and Idaho. This extensive lava plateau blankets more than two hundred thousand square miles, and in some places massive sheets of hardened basaltic lava are piled as much as four thousand feet thick. In short, much of Earth's more spectacular scenery has been created by the work of ancient volcanic activity and carved from igneous rocks. Thanks to Desmarest, von Buch, and other volcanologists, we now understand the meaning behind these landscapes.

Modern geologists concede that Werner's Neptunists were not 100 percent wrong: Some rocks actually *do* have an aqueous origin. These are characterized by the second great rock group: the *sedimentary rocks* that have formed from loose rock fragments called sediments. Some sediments, those derived from the decay and disintegration of previously existing rocks, have been moved from their point of origin by some agent of erosion. The wind' may pick up tiny dust-size particles and carry them halfway around the globe; rivers carry sand and silt to the sea; and glaciers can transport boulders the size of a house. But sooner or later, the transported sediments must be deposited. When this happens, the sediments are typically laid down in layers, one on top of the other. These layers or beds, called strata, are the most distinctive features of sedimentary rocks. Sandstone, clay shale, and conglomerate (a coarse rock composed of more or less rounded pebbles of assorted sizes) are typical examples of stratified sedimentary rocks. Other sedimentary rocks have—as Werner suspected—been precipitated from solution in sea water. Such common inorganic chemical sediments as rock salt, gypsum, and certain kinds of limestone have formed in this way. Other sedimentary rocks, coal and most fossiliferous limestones, for example, are organic in that they consist of the remains or products of ancient plants and animals.

In contrast to igneous rocks, which may form in the deep interior or on Earth's surface, sedimentary rocks are created by geologic processes that operate in the surface or near-surface, rock-forming environments. Thus, sedimentary rocks comprise only about 5 percent of the outer ten miles of the crust. Far more important, however, is the fact that these rocks make up almost 75 percent of the rocks that are exposed on the earth's surface. These rocks not only weather to form soils, they are an important source of many valuable products such as water, petroleum, salt, and other mineral resources. In addition, sedimentary rocks are of special interest to the historical geologist—they are the rocky pages of earth history that reveal much of our planet's past.

The third, and most complex, group of rocks are formed

in only one environment: deep within the earth. These are the *metamorphic rocks,* which are produced by processes that change preexisting rocks (igneous, sedimentary, or metamorphic) into totally different rock types. Thus, the process of metamorphism (Greek *meta,* "change," plus *morphe,* "form" or "shape") may transform a fine-grained limestone (a sedimentary rock) into a harder, coarse-textured, crystalline, metamorphic rock called marble. The effects of metamorphism are largely controlled by the chemical and physical properties of the original rock and by the agent and degree of metamorphism that is involved. But typically, the more basic changes are in the texture and chemical composition of the rock.

What geologic processes are capable of producing such drastic physical change? Rock can be metamorphosed in a number of ways. During mountain-building movements and other major crustal disturbances, tremendous forces may squeeze, bend, break, and otherwise deform the rocks that are involved. This in turn may generate sufficient heat and pressure to cause minerals in the rock to become more compact, forming a tightly interlocking mass. Rock may also be invaded by mineral-bearing gases and liquids boiling up from nearby magmas. As these seep into the surrounding rock, they may dissolve some of the original minerals and leave new ones in their place. In this way new minerals—some of great value—may form in rocks that were originally of no economic importance. As would be expected, metamorphic rocks are common in those areas that have been severely deformed by extensive crustal movements and/or baked by igneous intrusions. And, because they have had time to undergo such great change, metamorphic rocks are usually very old. Although not so abundant as igneous or sedimentary rocks, metamorphic rocks normally have considerable geologic significance. Because of the conditions under which they formed, these "made-over" rocks have recorded some of the more violent chapters in the history of the earth.

Anatomy of the Earth

Recent views from space have clearly shown that some areas of the Earth look smooth, while others appear to be quite rough and irregular. The uneven patches on Earth mark the continents, or landmasses, which constitute approximately 40 percent of the Earth's surface. Composed largely of granite, these rocky platforms have an average elevation of about three miles above the floors of the surrounding ocean basins. But the view from space is a bit misleading, for we can only see the 29 percent (57.5 million square miles) of the land that rises above sea level. Not so noticeable are the submerged continental shelves —the drowned margins of the continents that are awash by the sea. These gently sloping extensions of the land make up the remaining 11 percent of the continents.

The surface of the continents is quite irregular, consisting of varied land forms such as plains, plateaus, and lofty chains of mountains. In elevation, these range from coastal plains that touch the ocean's waters to the summit of Mount Everest, which towers 29,028 feet above sea level.

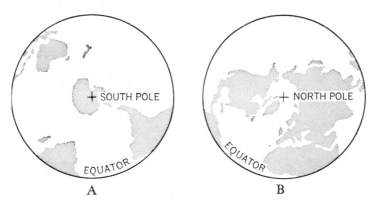

FIGURE 7   The Southern Hemisphere (A) has been called the water hemisphere because most of Earth's landmasses are located north of the equator (B).

But on the average, the continental surface lies only about one-half mile above the sea.

The view from space clearly indicates that the smooth areas greatly outnumber the bumpy continental land-masses. This is especially obvious on space photographs taken of Earth's Southern Hemisphere, for more than 81 percent of this part of the earth is covered by water. In marked contrast, views from above the North Pole show that the continents cover almost 40 percent of the area north of the equator. The over-all, or composite view, of both spheres reveals that almost 71 percent—70.8 percent, to be specific—of the lithosphere's surface is hidden beneath the sea. This represents an area of 139 million square miles.

But these smooth areas are somewhat deceptive, for beneath them lies the rugged terrain of the ocean floor. Originally thought to be relatively flat and featureless, recent oceanographic surveys have shown that the sea-covered part of the lithosphere is every bit as diversified as that which lies above the sea. Consider, for example, the Mid-Atlantic Ridge. This great underwater mountain chain averages about twenty miles in width, rises roughly a mile above the ocean floor, and zigzags for some ten thousand miles from Iceland to the southern tip of Africa. In many places the monotony of the sea bottom is broken by deep canyons and submarine trenches that dwarf many similar chasms on land. The deepest of these, the Pacific Ocean's Mindanao Trench, plunges more than 37,700 feet below sea level—some seven times deeper than Arizona's famed Grand Canyon. A similar "deep," the 30,246-foot Puerto Rico Trench, marks the deepest part of the Atlantic Ocean. In between these extremes are plains, plateaus, and smaller elevations and depressions of all shapes and size. Yet despite the many surveys that have been made to study the ocean floors, only about 6 percent of this part of the lithosphere has been mapped with any degree of accuracy. It is small wonder that it is often said that we know more about the moon than we know about the surface of the earth.

Although the surfaces of the continents and ocean basins

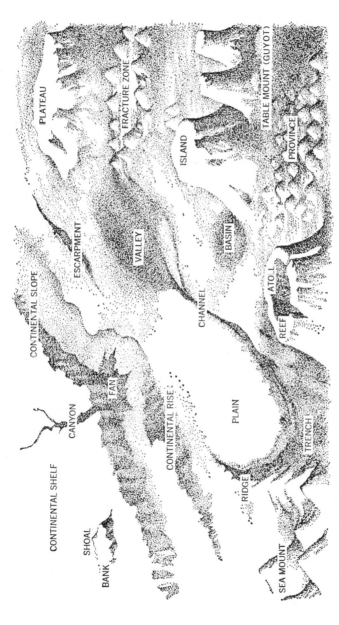

FIGURE 8 Submarine landforms such as these break the monotony of the vast ocean floor.

are similar in relief, they differ considerably in composition. As noted above, the landmasses are underlain by granite —an igneous rock with a density of about 2.7. In places the sea bottom is covered by layers of marine sediments, but a heavier rock called basalt underlies the floor of all ocean basins. This rather dark rock has a density of about 3.3, and geophysical studies indicate that the lighter granitic continents are literally "floating" on the heavier basaltic "sea." Geologists consider the continents and ocean basins to be the two major surface features of the earth. And although there is more than twelve miles difference between the highest and lowest places on Earth's surface, the lithosphere's irregularities are inconsequential in comparison to the size of the earth. As noted earlier, on a model Earth of twenty-five feet in diameter, Mount Everest would rise less than a quarter-inch above sea level.

Much has been learned about the surface of the solid earth, but what lies underneath the thin, rocky veneer beneath our feet? Jules Verne's *Journey to the Center of the Earth* notwithstanding, man has only ventured about two miles into the earth's interior: in a deep gold mine in South Africa. And even in this day of advanced technology, he has only drilled some five miles into the earth. This penetration—in an unsuccessful oil test in West Texas—is but a fraction of the four-thousand-mile distance to the heart of the earth.

But though man cannot personally journey into the bowels of our planet, he has still learned much about the inner earth. Most of this information has come from indirect evidence obtained from field observations made on certain rocks exposed on the surface and from laboratory experiments carried out on these surface rocks and minerals. Other studies based on Earth's magnetic field and the flow of heat from within the earth have provided additional bits of information. Oddly enough, clues as to the nature of Earth's interior have even come from outer space. These data have been provided by meteorites—which might be fragments of a former planet—and by comparing Earth with celestial objects such as stars, the sun, and other planets in our solar system.

However, most of what we know about "inner space" has been learned from the study of seismic (earthquake) waves in conjunction with laboratory experiments. These investigations reveal that seismic waves travel at different speeds through materials of different densities, or specific gravities. The waves travel faster in very dense rock than they do in lighter rock. Thus, by measuring the time that it takes a certain type of wave to travel from an earthquake's origin to distant seismograph stations, it is possible to estimate the density of the materials of the earth's interior. It has also been learned that seismic waves bend or change direction when they reach certain depths in the earth. The depths at which these waves are deflected correspond with changes in the densities of the rocks. Seismologists have used this information to construct a mental model of what may be inside the earth, and this is used as the standard model of Earth's interior. What is this model like? The lithosphere seems to be divided into three rather distinct zones: the crust, the mantle, and the core.

Studies of volcanoes led early geologists to believe that Earth's interior was full of molten rock. They also thought that Earth's surface rocks represented a solid, rocky rind that was floating on this liquid interior. Although this idea has long since been abandoned, the word "crust" is still used to refer to the rigid outer shell of the earth. Compared to the rest of the lithosphere, the crust is very thin and comprises but a minute fraction of Earth's total volume. In volume, the crust makes up only about 1 percent of Earth's 259,930,000,000 cubic miles. Its thickness ranges from as little as three miles in certain ocean basins, to as much as thirty miles beneath the higher mountains. Not all crustal rocks are alike. Those that underlie the ocean basins are heavier than those that lie beneath the continents, for the oceanic crust consists primarily of dense basaltic rocks similar to those commonly erupted from volcanoes. The continental crust is somewhat more complex. It appears to consist of two rather distinct layers: an upper zone of dense, light-colored granitic rocks and a lower layer of dark, somewhat heavier, basaltic rocks similar to those of the oceanic crust.

After passing through the crust, seismic waves abruptly increase their speeds. This marks the boundary between the relatively dense lower crust and the mantle, an eighteen-hundred-mile-thick zone of very dense rock. The point at which the seismic waves suddenly speed up was first noted in 1909 by Andrija Mohorovičić, a Yugoslav seismologist, and is called the Mohorovičić Discontinuity or, more simply, the Moho. The Moho should not be confused with Project Mohole, now-defunct, which was scheduled to drill through the crust and obtain samples of the mantle. Hopefully, Project Mohole may be reactivated at some later date.

The core of the earth is about 4350 miles in diameter and is surrounded by the mantle. Although comprising only 15 percent of Earth's volume, the core's heavy materials make up about one-third of the planet's total mass. The core apparently consists of two parts: an exterior, probably liquid, outer core, and an inner core that is believed to be solid. The twofold nature of the core is based on studies of the behavior of earthquake waves as they reach the center of the earth. Seismic waves capable of passing through the 1360-mile-thick outer core are greatly slowed down and behave as if they were passing through a liquid. However, upon reaching the inner core, the velocity of the seismic waves suddenly increases, and they act as if they were again passing through a solid. More important, one type of wave—which will not pass through a liquid—is not transmitted through the core at all. This strongly suggests that the outer core is liquid and the inner core is a very dense solid. How is it possible for both solid and liquid material to be in the core? The inner core is thought to remain in solid state because it is at a high pressure (even though also at a higher temperature) than the outer core that surrounds it. Though seismic studies have provided clues as to the nature of the core, its composition is not definitely known. Nevertheless, there is reason to believe that it consists of 80 to 85 percent iron with varying amounts of nickel, silicon, and cobalt.

Thus, from such varied sources as photographs taken from space, sea-floor soundings made by oceanographers, seismic "messages" from inner space, and the worldwide studies of countless geologists, we have learned much about the solid earth and its interior. Even so, questions about the earth still far exceed the answers, and many mysteries are still locked deep within this planet.

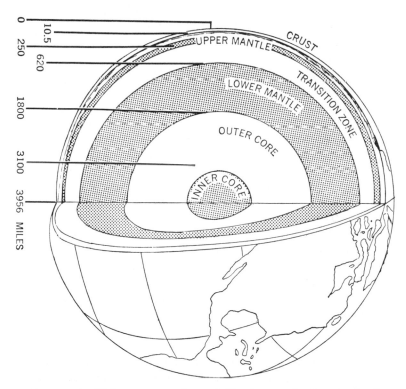

FIGURE 9 This cutaway view of Earth reveals the standard "model" of the earth's interior.

6

THE DYNAMIC EARTH

"Good old terra firma," "the everlasting hills," "solid as a rock." These rather common sayings are frequently used to emphasize stability and permanence, for they assume the earth to be the ultimate symbol of all that is stable and dependable. But is it? Many residents of California and other earthquake- and landslide-prone areas will quickly assure you that good old "terra" is not always as "firma" as one might suspect. Nor are the hills "everlasting." Earth's loftiest mountains are perpetually being gnawed away by the relentless action of running water, weather, ice, and gravity. Given enough time—and there is always plenty of time—even the mightiest mountain range will eventually succumb and be reduced to sea level. The solid rock? Placed in the proper environment, *any* rock can be squeezed until it will fold and bend like soft plastic. In short, Earth is not the inert, static globe that it was once thought to be. Rather, it is a dynamic and rather tightly "wound" planet that is full of energy and constantly changing both inside and out.

Evidence of the dynamic earth can be seen the world around. Volcanoes are erupting, the land is jolted by earthquakes, giant sea waves smash into the shore, and once-mighty glaciers are slowly melting away. Less dramatic, but far more significant, are the more subtle changes in the earth: weathering to produce soil, the erosive work of streams, and the slow accumulation and sinking of sediments that will become the rocks of the geologic future. Indeed, we might describe Earth's surface as a battleground where two sets of opposing forces have been locked in timeless combat. On one side are the internal forces that originate deep within the earth. These powerful forces have elevated ancient sea floors many miles above sea level, torn apart great segments of the earth's crust, and covered parts of the world with thick layers of lava and volcanic ash. The inner forces that produce such great change do not go undefied. Rather, these internal processes are continually challenged by external forces that have their origin at or near Earth's surface. Although we are not always aware of this never-ending conflict, we should be thankful for it. Without the constant interaction of these two opposing forces and the mechanism that maintains the delicate balance between them, we would not be here today.

The Delicate Balance

We have seen that the earth's major relief features consist of continents and ocean basins. It is also known that the ocean basins and landmasses have not always been exactly as they are today. In ages past, internal forces have raised great segments of the earth's crust to form areas of high relief. And no sooner had these regions been uplifted than the external agents of erosion began to whittle them down. Thus, throughout geologic time mountain ranges have been worn away and the materials eroded from them and deposited elsewhere. These alternating periods of uplift and erosion continually redistribute the earth's surface materials, and the repeated shifting of the crustal load upsets Earth's gravitational equilibrium.

What will be the outcome of this terrestrial battle of "ups and downs"? Thanks to the condition known as isostatic compensation, it will probably always remain a draw. The principle of *isostasy* (literally, "equal-standing") assumes that at considerable depth within the Earth, different segments of the crust will be in equilibrium with other sections of unequal thickness. The differences in height of these crustal segments—continents and ocean basins, for example—are explained as the result of differences in density of the rocks. Consequently, the continents and mountainous areas are high because they are composed of lighter rocks; the ocean basins are lower because they consist of much denser (heavier) rocks (FIGURE 10).

As noted earlier, seismic waves indicate that the light crustal rocks beneath mountains extend more deeply into the mantle than do continental areas of low elevation. In other words, mountains literally have "roots" that project

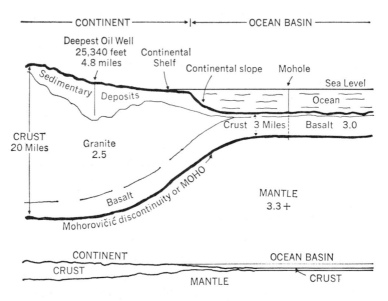

FIGURE 10   By permission from *Geology: An Introduction* by R. L. Bates and W. C. Sweet, D. C. Heath & Company (1966). Cross section about four hundred miles long through the earth's crust. Numbers refer to relative density. In order to show detail, the vertical scale of the section has been exaggerated ten times. In the lower sketch the section is drawn to true scale.

downward into the earth. This is somewhat analogous to an iceberg that rises, say, one hundred feet above the ocean surface, but extends downward for several hundred feet beneath the sea. When the iceberg melts, the base of the iceberg will rise as the top melts away. By the same token, when material is eroded from the top of a mountain range, this segment of the crust will rise for the same reason that the base of the iceberg rises. On the other hand, when the sediments removed from the mountains are laid down in a depositional basin, this part of the crust will sag because of the additional weight that has been added there. As the continents are eroded and sediments deposited in the sea, the ocean basins are depressed. The downward pressure exerted by those accumulating sediments causes displacement of the plastic subcrustal rocks, which tend to push the landmasses up. The rise of the continents is also given an assist by erosion, which removes rock material from the land, thereby making the continents lighter and more susceptible to uplift.

Because the movements of isostatic adjustment are essentially vertical, isostasy cannot be used to explain the horizontal compressive forces that fold rocks into mountain ranges. However, it does explain the delicate state of equilibrium that keeps the Earth's landmasses from being completely eroded to a continuous level surface.

The Forces from Within

Oregon's Crater Lake National Park is generally considered to be one of the world's most scenic areas. Its unique geologic setting, the lake's incredibly blue water, and the varied fauna and flora blend together to produce one of the most peaceful landscapes on Earth. Equally beautiful is mighty Mount Rainier, whose ice-clad flanks support a glistening system of Alpine glaciers. But the present serenity of these two popular national parks is not in keeping with their geologic history. The deep basin occupied by Crater Lake is the product of terrific volcanic explosions and the thunderous collapse of a mountaintop. Mount Rainier has an equally fiery past—this volcanic cone is built of alternating layers of lava and volcanic ash thrown out during countless explosive eruptions.

Consider also California's magnificent Sierra Nevada and the less lofty, but equally beautiful Appalachian Mountains. These great mountain ranges account for some of our nation's most lovely scenery, but they, too, were born of violence. The Sierra Nevada consists of a broken block of the earth's crust that has been tilted and elevated with respect to the surrounding land. The Appalachians, on the other hand, are the result of powerful internal forces that buckled the earth's crust, folding it into an immense series of "wrinkles" of rock that bulge upward.

Yet despite the diversity and wide geographic distribution of these four scenic areas, they all have one thing in common: Each of them is dramatic surface evidence of the internal forces at work deep within the earth. What forces are great enough to crumple rocks to create a mountain range or cause a mountain to explode? Crater Lake and Mount Rainier are products of *volcanism*—the natural processes that result in the formation of volcanoes and volcanic rocks. The Sierra Nevada and Appalachians have been produced by the forces of *diastrophism*. This process, known also as tectonism, has produced large-scale deformation in the earth's crust and is responsible for most of our great mountain chains.

Volcanic activity has fascinated man since the beginning of recorded history. Volcanoes are equally interesting today and—though they have been extensively studied for many years—geologists still have not solved all their mysteries. The biggest problem, of course, is simply: Why do we have volcanoes? As noted earlier, we know that molten rock material may form at various places within the earth's crust. These mixtures of melted minerals are typically found in isolated "pockets," or magma reservoirs. When this molten rock spills out on the surface it is called lava,

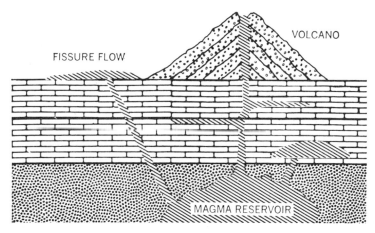

FIGURE 11 By permission from *The Story of the Earth* by William H. Matthews III, Harvey House, Inc., Irvington-on-Hudson, N.Y. (1968). Cross section of a volcano showing magma reservoir, cone, and nearby lava flow.

and when lava cools and hardens it may produce a variety of volcanic rocks. The rock formed in this way is also called lava. The molten rock rising from within the earth has no connection with Earth's outer liquid core. There is no underground "pipeline" that transports magma from the core, as was once suggested. The core is more than eighteen hundred miles beneath the surface, whereas volcanoes are thought to originate in the earth's crust and upper mantle—not more than twenty miles within the earth.

What causes these rocks to melt? Although several theories have been advanced to explain the origin of volcanic heat, volcanologists are still not sure of its source. They all agree, however, that incredibly high temperatures are required to convert solid rock to the liquid state. It is also apparent that melting of the rock occurs at depths where earth materials are subjected to very great pressure and intense heat. Now, in order to melt, rock must expand. And, because the deeper subsurface rocks are tightly compressed by the rocks above them, more heat is required to melt these rocks than to melt the same type of rock nearer the surface. In other words, the temperature at which rocks will melt becomes higher as pressure on the rocks becomes greater. Thus, it requires more heat to melt rocks that are being compressed by the burden of several miles of overlying rocks than to melt similar rocks exposed on the earth's surface.

This has led some earth scientists to believe that the more deeply buried rocks may be in a liquid rather than solid state. They reason that this is possible because temperature measurements taken in deep mine shafts and bore holes show that there is a steady temperature increase as one goes deeper into the earth's crust. The rate of increase is approximately 1°F for each sixty feet of depth, or roughly 150°F per mile. This increase is known to continue to a depth of about five miles, which is as far as measurements have been made. If it can be assumed that subterranean temperatures continue to increase at this rate, temperatures that would exceed the melting point of all known rocks would eventually be reached. Could these deep-seated temperatures account for the pockets of liquefied rock? Probably not. Data from seismic waves clearly show that—with the exception of isolated zones of igneous activity—most rocks of the crust and upper mantle are quite solid. This suggests, then, that the rise in subsurface temperatures does not continue at a constant rate of increase. In short, while higher temperatures may have contributed to the melting of deeply buried rocks, some other heat source must also be accounted for.

One source of additional heat may have been the result of friction generated by powerful movements such as earthquakes and mountain building. It is reasonable to assume that a great deal of heat is produced when huge masses of solid rock fold, break, and slide over one another. Moreover, this idea is supported by the fact that most areas of volcanic activity are located in or near regions of recent earthquake movement and mountain building. These great crustal disturbances normally cause arching and fracturing of the crust, and such deformation might locally reduce the pressure on more deeply buried rocks. This pressure reduction would permit the rocks to expand, thereby allowing them to melt at somewhat lower temperatures.

Some geologists have suggested that the necessary heat may come from concentrations of radioactive elements in the earth's crust. These energy-producing elements, such as uranium, are capable of generating much heat, and are known to be relatively abundant in certain crustal rocks. Deeply buried pockets of highly radioactive material could conceivably produce sufficient heat to liquefy the rocks enclosing them. As the rocks melted they would expand and fracture the overlying rocks. These fractures would then provide zones of weakness that would permit further expansion and through which the magma might ultimately reach the surface.

But in the final analysis, it seems that a combination of factors produce the temperatures that cause rock to melt. Some of the heat comes from the high temperatures inherent in deeply buried rocks. The rest is derived from radioactivity with, perhaps, an occasional assist from friction-produced heat associated with crustal deformation.

Most of us know what happens when molten rock reaches the earth's surface. Volcanic eruptions are periodically in the news, and the sight of molten rock, along with showers, ashes and cinders, spewing out of a volcano is not rare on television and in the newspapers. The volcanic debris from such eruptions commonly builds a cone-shaped mountain, and volcanic material may continue to be erupted through its central crater. Some, such as the volcanic mountains that form the Hawaiian Islands, rise tens of thousands of feet from the ocean floor. Countless others are widely scattered on many parts of the landmasses. Many of these, Mount Rainier, Mount Shasta, and Mount Fuji, for example, are among the world's most famous and beautiful mountains.

Not all volcanic activity produces the typical cone-shaped volcanic mountain. Lava may also simply pour out on the surface and spread over the land to form extensive lava flows like those that cover the Columbia River Plateau of Oregon, Idaho, and Washington. This great lava plateau consists of horizontal layers of basaltic lava that escaped through great fissures in the crust, and eventually spread in sheetlike masses over an area of some two hundred thousand square miles. In places the flows produced by this prehistoric lava flood are as much as four thousand feet thick.

Crater Lake is especially unique, for it is a rather interesting byproduct of volcanism. Many thousands of years ago the location of Crater Lake was the top of a huge volcano known as Mount Mazama; this volcanic cone was part of the Cascade Range and was probably about twelve thousand feet in height—almost as high as Mount Rainier. Where is mighty Mount Mazama today? Oddly enough, most of the volcano's top is at the bottom of Crater Lake. About six thousand years ago, a series of great explosions ripped out much of Mazama's interior. These violent eruptions threw out great quantities of ash, lava, and cinders, and soon drained the lava from beneath the volcano and weakened the upper part of the mountain. Without the underlying support of the erupted material, Mount Mazama simply caved in under its own weight. The collapse of the mountaintop produced a great bowl-shaped depression called a *caldera,* and as time passed the caldera

FIGURE 12   Reproduced from *Crater Lake: The Story of Its Origin* by permission of Howel Williams and the University of California Press. The evolution of Crater Lake. (A) Beginning of the great eruptions. (B) Eruptions become more violent and pumice showers heavier. Lava level in pipe is falling. (C) Climax of eruptions. Glowing avalanches sweep down the sides of the volcano. Magma chamber is being rapidly drained. (D) Summit collapses into magma chamber. Gas vents appear in the caldera floor. (E) Crater Lake today. Wizard Island and lava are shown on the lake floor. Magma in underlying chamber is in large part, or entirely, solidified.

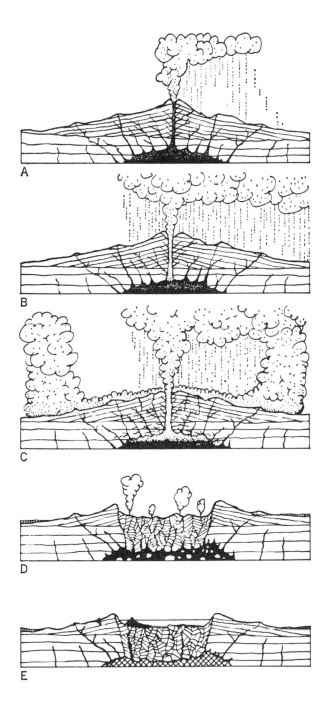

A

B

C

D

E

became filled with water, and the change from volcano to lake was complete. But geologists know that old Mazama did not die easily. As recently as one thousand years ago, a series of vents opened up on the caldera floor. These openings must have been the source of explosive eruptions, for they produced three cinder cones. One of these—Wizard Island—can still be seen in the lake today.

Diastrophism—the process responsible for the Appalachians and the Sierra Nevada—is the other major force at work within the earth. Although they produce some spectacular landforms, diastrophic movements generally occur very slowly, for great mountain ranges do not literally "pop" out of the earth overnight. Instead, *orogenic* (mountain-building) processes operate over vast spans of time, and their final product may be the culmination of tens of millions of years of crustal adjustment. Even less noticeable are the results of slow vertical movements that have rather gradually brought about broad uplift or submergence of the continents. Rock strata involved in this type of movement are not usually as greatly deformed as in horizontal or more sudden diastrophic changes.

But though the forces of diastrophism work very slowly, they can, and frequently do, culminate quite unexpectedly and disasterously. This happened on March 27, 1964, a day that most residents of Alaska will never forget. It was at 5:36 P.M. on that Good Friday when south-central Alaska was wracked by an exceptionally devastating earthquake. This quake, like most, did not last long—probably only about four minutes. But during those 240 seconds more than 114 people were killed and almost $750,000,000 worth of property was destroyed.

Earthquakes, the most familiar sudden diastrophic movement, are common in certain parts of the world, and locally may be the cause of great loss of life and property damage. Nevertheless, the crustal displacement caused by even the most severe earthquakes has been measured in tens of feet, rather than thousands of feet as in the case of orogenic movements. The process of *faulting*—the fracturing and displacement of rock—is responsible for most earthquakes. Faulting also plays an important role in mountain building, as evidenced by the mighty Sierra Nevada, for this fault-block mountain range consists of a massive segment of rock that broke loose in the earth's crust. The rock mass that forms the Sierra Nevada is one of the largest fault blocks known. Ranging from forty to eighty miles in width and more than four hundred miles long, this single, tilted block forms an unbroken mountain chain almost as extensive as the Swiss, French, and Italian Alps. The eastern front of the block juts more than eleven thousand feet above the valley to the east of it. Its broad, more gently sloping west flank dips into the Great Valley of California and thence under the Pacific Ocean. Thus, the abrupt eastern margin, where the faulting occurred, rises more than two miles above sea level, and its western edge lies almost five miles beneath the ocean.

Needless to say, prodigious force was required to displace this massive block of earth material. Where does such

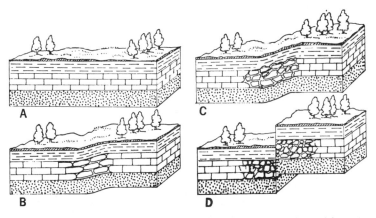

FIGURE 13 A normal section of rocks (A) may be subjected to such great stress (B and C) that it may eventually break (D), thus causing an earthquake. (Reproduced by permission from *The Story of the Earth* by William H. Matthews III, Harvey House, Inc., Irvington-on-Hudson, N.Y.)

energy come from? As in volcanism, the causes of faulting
—and the earthquakes it spawns—are not completely understood. It is well known, however, that the crustal rocks are
commonly subjected to great stress and strain, and some of
these pressures may be maintained for thousands, even
millions, of years. As these rock masses undergo prolonged
pressures from opposing directions, they slowly bend and
change shape. Continued pressures may produce such great
strain that the rocks will eventually break and then suddenly snap back into their original unstrained state. It is
this "snapping back," or *elastic rebound,* that displaces
fault blocks and generates the wavelike movements in the
rocks that cause earthquakes. Earthquakes have probably
jarred this planet since time began. And they still do, for
each year more than one million earthquakes are recorded.
Fortunately, only about seven hundred of these annual
shocks are strong enough to do appreciable damage or
cause loss of life. One reason for this is because most earthquakes occur beneath the sea. Even so, hardly a month
passes that we do not read of the devastating effects of
these sudden diastrophic movements.

By far the most striking evidence of diastrophism can be
seen in the distorted rock formations of great folded mountain chains such as the Appalachians, Alps, and Rocky
Mountains. These ranges have not only been elevated
high above sea level, they have also been subjected to
powerful lateral or horizontal compressive forces. Forces of
this type are more localized, and operate essentially parallel to the earth's surface, as opposed to the vertical movements that produce the broader and less obvious elevation
and submergence of the continents.

The nature of the lateral crustal adjustments that produce folded mountain ranges have long puzzled geologists.

FIGURE 14   Mountains are not alike in form or origin. (A)
Mountains formed by volcanic action; (B) mountains resulting from folded layers of rock; (C) mountains formed from
fault blocks; (D) mountains formed as a result of vertical uplift.
(U. S. Geological Survey)

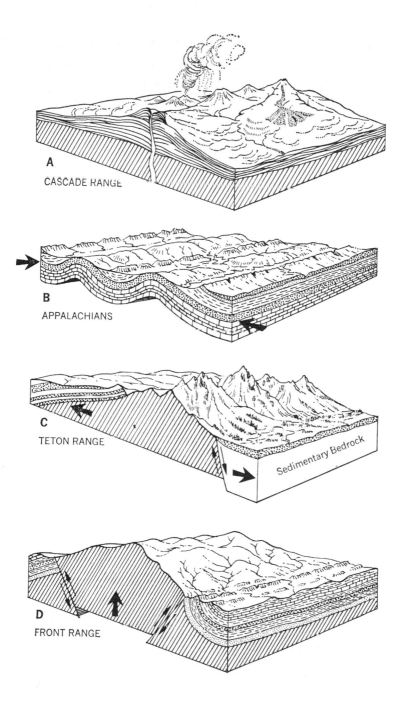

A CASCADE RANGE

B APPALACHIANS

C TETON RANGE

Sedimentary Bedrock

D FRONT RANGE

To explain the origin of these pressures and other internal phenomena such as volcanism and earthquakes, a number of theories have been proposed. But as might be expected, none of these explanations appear to have all of the answers, and serious objections have been raised against each. Space precludes a detailed discussion of the pros and cons of these hypotheses. Nor is this *Invitation to Geology* meant to provide in-depth treatment of such subjects. However, a brief review of these major theories can be helpful to an understanding of basic geologic phenomena and the geological insight of earth scientists. More important, they emphasize some of the problems encountered by earth scientists as they attempt to provide explanations for some of the more basic forces at work within the earth. These hypotheses are also typical examples of how the geoscientist uses indirect evidence and deductive reasoning to make logical speculations about phenomena that cannot be directly observed.

One of the oldest of these hypotheses is the so-called Contraction Theory, which assumes that the earth is cooling from an original molten condition. As the interior of the earth cooled, the planet contracted, and this produced compressional stresses that caused the rocks of the outer crust to become crumpled and wrinkled. This hypothetical condition might be compared to the wrinkling of an apple's skin as its interior shrinks during drying. In other words, the "wrinkles" in Earth's rocky rind are analogous to those in the peel of the dried apple. The Expansion Theory assumes just the opposite, for it supposes an original Earth of about one-half its present diameter of seventy-nine hundred miles. The granitic crust is claimed to have uniformly covered this "miniplanet," but subsequent expansion caused faults that produced crustal blocks that became the present continents. Continued expansion caused these blocks to become widely separated, and the space between them became the ocean basins.

One of the more recent proposals is the Convection Theory, which suggests that thermal convection currents beneath the crust may cause the rocks to expand and push upward. Currents of this type could produce massive upwelling of earth material, which might be compared to movements produced in a pot of boiling oatmeal. Such currents are generated in liquids and gases by heating, for when heat is applied to the cooler portion of a liquid it sinks because it is more dense than the heated portion. This pushes up the heated part, which then loses its heat upon reaching the surface. As this liquid cools, it becomes more dense and gradually moves downward again. This continual exchange of heat sets up circulating currents of convection cells as long as heat is applied. It is argued that giant convection cells in the mantle could account for Earth's mountain belts as well as certain other structural features (FIGURE 15). And, because the mantle is under great heat and pressure, the rocks in this part of the earth would behave as a very viscous material. Frictional drag between the crust and rock flowage in the mantle would produce the necessary force to cause crustal displacement. What produces the heat necessary to energize these convection cells? It is apparently generated by the steady decay of radioactive elements. Because the overlying rocks conduct heat very slowly, there is a steady buildup of heat that causes the material in the mantle to expand and produce plastic flow upward. Within recent years many geologists have come to favor a combination theory, which assumes that the crust is expanding as the earth cracks along fracture lines. It is assumed that molten rock material would well up along these fractures, pushing the broken crustal segments farther apart. This would subject the crustal blocks to tremendous pressures, which would cause the rocks to buckle upward. The Theory of Continental Drift discussed in Chapter 8 has also been used to explain the origin of folded mountain ranges. Supporters of this theory believe that Earth's great belts of folded mountains may have been caused by crumpling of the landmasses as they drifted along on the plastic rocks of the mantle.

Thus, two powerful—and often antagonistic—forces have

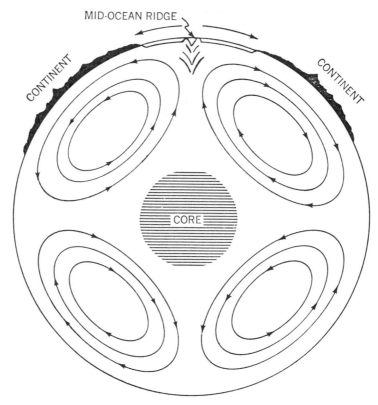

FIGURE 15　The Convection Theory holds that mountain ranges and midocean ridges form when convection currents in the mantle cause the rocks to push upward.

played vital roles in the development of the earth. And although a vast amount of research and scientific speculation has been directed toward a better understanding of volcanism and diastrophism, there are still many questions to be answered. But despite these unsolved problems, much has been learned about the nature of these two great internal forces. Geologists have learned that most of Earth's surface features and many internal phenomena can be explained in terms of some phase of these basic geologic processes. They have also found that these forces have had a hand in forming some of our more important natural resources and shaping the history of the earth and its inhabitants.

Forces That Shape Earth's Face

Although not always so obvious or dramatic as changes produced by volcanism and diastrophism, the external forces that alter the earth's surface are generally of greater importance to man. These changes—most of which operate so slowly and subtly as to be imperceptible—are taking place all around us. But given enough time, the products of their handiwork become readily noticeable. Evidence of the work of surface geologic processes is so common that it is generally taken for granted, and many of the processes operate under our very noses. Rocks are slowly destroyed and mountains gradually lowered as rain and frost attack the minerals of which they are made. Each year the earth's surface streams erode millions of tons of rock fragments from the land and deposit them in the sea. And along the shore, ocean currents and waves rearrange and redeposit some of this material to form beaches and sand bars. But much of this rock debris accumulates on the ocean floor and will later become the sedimentary rock of the geologic future. In some parts of the world, the restless desert sands are continually shifting in response to the wind. Elsewhere powerful rivers of ice gradually plow down the sides of mountains, gouging out valleys and carrying along great quantities of rock to be deposited when the glacier melts. In short, the surface materials of the earth are continually being altered and rearranged by the geologic agents that operate on or near its surface.

Unlike the internal geologic forces that cannot be studied by direct observation, the geologist can often see the external geologic agents at work. As a result, he generally knows more about them and the ways in which they operate. But though the internal and external forces differ in many respects and frequently are in opposition, they have one thing in common: Both require tremendous amounts of energy. Where does this energy come from? By and large, most processes that operate on the earth's surface derive their energy from the sun. Our parent star radiates extraordinary amounts of energy into space, and although only a fraction of this energy reaches our planet, it is sufficient to warm the air, the oceans, and the solid earth. The transfer of energy within and between these three spheres results in the continual distribution of Earth's surface materials and the shaping of its surface features. But the sun's energy is not the only force at work on the earth's surface. The earth's gravitational field also plays an important role in the operation of external geologic processes. Solar energy and gravity vitally affect biological and geological processes, and our lives literally depend upon them. However, considered within the context of how these forces energize external geologic processes, we shall concentrate on the transfer of energy within and between the atmosphere, hydrosphere, and lithosphere.

Leonardo da Vinci, the remarkable artist-scientist-engineer, once said: "The air moves like a river, carrying the clouds with it." This observation—made four hundred years ago—was astonishingly correct, for the air literally does carry "rivers of water." By means of solar heating, water is evaporated from the sea (the basic source of atmospheric water) and condenses in the air as clouds of water vapor. Powered by the sun's energy, winds transport these water-bearing clouds to the continents. Atmospheric water is eventually released by precipitation in the form of rain or snow, and upon being liberated from the atmosphere, water comes under the influence of the earth's gravitational field. Some water falls directly back into the ocean, but much of it falls on the land, where rivers and underground flow may eventually return it to the sea. This system of water circulation from oceans to atmosphere and back to oceans, either directly or via the land, is called the

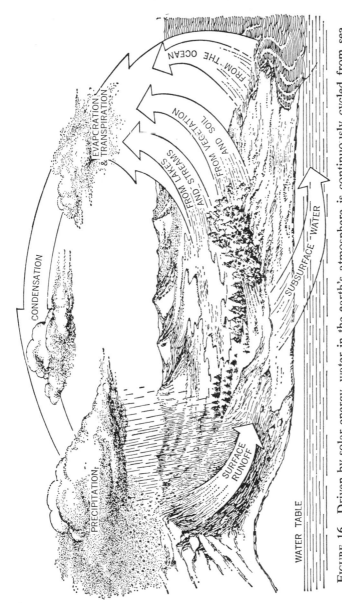

FIGURE 16 Driven by solar energy, water in the earth's atmosphere is continuously cycled from sea to land and back to sea. (U. S. Geological Survey)

hydrologic cycle. Driven by solar energy and the force of gravity, the hydrologic cycle is a natural machine that is forever pumping, distilling, and moving the waters of the earth. These are essentially the same waters that have been recycled for more than four billion years, for little has been added or lost since the first clouds formed and the first rains fell on our then-desolate planet. This same water has been in perpetual movement from the sea into the air, dropped on the land, and funneled back into the oceans. The water cycle has neither beginning nor end, and during its constant operation it has played a most important role in sculpturing the landscapes of the earth's surface. Indeed, it is the unifying thread that binds the various geologic agents together.

Water returning to the sea by means of streams is a most effective geologic tool, for stream erosion has done more to shape the landscape than all other agents combined. The Grand Canyon of the Colorado River is a classic example of stream erosion. Located in northern Arizona, this 217-mile-long, nine-mile-wide canyon is almost one mile deep. This magnificent gorge has been carved by the Colorado River—with relatively minor help from other geologic agents—in the last seven million years: a comparatively short time, geologically speaking. But even as streams wear the land away, they carry the eroded particles of rock to other areas, where they will ultimately be deposited as sediments. There, much later in geologic time, they may be converted into sedimentary rock.

Excessive runoff causing continued denudation of the land can lead to destructive soil erosion and the loss of fertile topsoil. With man's increasing use—and abuse—the earth's surface becomes increasingly vulnerable to attack by running water and other surface agents. Consequently, he must continually develop new and more effective ways to save our life-giving soil.

If rocks are continually exposed to wind, rain, and ice, they gradually disintegrate and undergo chemical change. Rock weathering occurs wherever the atmosphere is in contact with the lithosphere and, although it operates very slowly, weathering has played an important role in the shaping of surface land forms. Continued exposure to atmospheric agents will cause rocks to be dissolved by rain water, pried apart by frost and ice, and scratched by wind-blown sand. Many of these weather-produced changes are purely physical and merely reduce the original rock to increasingly smaller fragments. But solution, oxidation, and other chemical changes can also cause the rock to decay and be altered chemically. Weathering, like all geologic processes, does not proceed at the same rate and on the same scale everywhere. Many factors will affect the rate at which rocks will weather, but the most important of these are climate, the elevation of the land, and the chemical and physical nature of the rock. Moreover, rates of weathering have varied greatly throughout geologic time because of changing climates, differences in protection provided by vegetation, and other factors.

Although disintegration and decomposition of the rocks do wear away the land and can cause many problems, weathering plays a vital part in biological and geological processes. Indeed, weathering—perhaps more than any one geological agent—serves to tie together Earth's four great spheres. Rock weathering comes about because of the interaction of the atmosphere, hydrosphere, and lithosphere. The end result of the weathering process is soil, and soil is the all-important "bridge" that connects the biosphere with three inorganic spheres of matter. So important is this relationship that earth and life scientists generally agree that there would be no life without soil and no soil without life—certainly not on the land.

At higher elevations in many parts of the world there are great moving "rivers" of ice called glaciers. Most of these ice masses originate in snowfields high in the mountains. Here the yearly snowfall and refreezing of melted snow exceed the over-all rate of melting, and the snow piles up in an ever-thickening blanket. As the snow pack becomes deeper, the lower layers are gradually squeezed together and converted into rounded grains of ice. With the passage of time and as additional snow is added on top of the granular ice, the lower portion of the ice pack gradually becomes pressed together to form solid glacial ice. Eventually the ice mass becomes so heavy that the lower layers begin to yield, the ice starts to move in response to gravity, and a glacier is born. But although the ice may flow down an abandoned stream channel, glaciers do not "run" like streams. Instead, they slowly "crawl" downslope at rates that are typically measured in inches per day or week.

As the glacier moves down its valley it greatly alters the rocks over which it passes. Some rocks become frozen into the ice and are plucked from the valley floor or wall; others fall on top of the glacier. These rocks may become embedded in the ice to form rocky "teeth" that gouge and scrape the rocks over which they ride. The glacier may also smooth and polish the rock with which it comes in contact. But although glaciers can carry great quantities of rock material, they must eventually deposit their loads. This commonly takes place when the glacier melts, and can result in a number of interesting land forms. The ice-carved peaks of the Tetons, Rocky Mountains, and Alps and beautiful Yosemite Valley are good examples of how glacial ice can sculpture the landscape.

In dry, desert, and semidesert areas, the wind can be a potent geologic force. Precious topsoil can be blown away and later deposited as sand dunes like those at White Sands National Monument in New Mexico. And along the seacoast, waves and currents relentlessly attack the land, eroding here and depositing there. Groundwater, which works just beneath the surface, can also be an effective geologic agent, for slowly percolating underground water has dissolved great caverns out of soluble rocks. The same caves have later been decorated with stalactites and stalagmites formed when the subsurface waters deposited their dissolved mineral load. Such is the origin of famous underground attractions like Carlsbad Caverns and Mammoth Cave.

In other areas, gravity may cause great masses of rock and mud to move downslope. The pull of gravity is the force behind landslides like the one that clogged Montana's Madison River Canyon in the Hebgen Lake earthquake of August 1959. Triggered by the earthquake, this slide lasted only a few moments, but during that time some forty million cubic yards of rock were jarred loose from the canyon wall. Moving at an estimated speed of one hundred miles per hour, this mass of rock—weighing at least eighty million tons—crashed down across the valley and rode four hundred feet up the opposite side of the canyon. When it was all over, the valley was filled with a sheet of rock debris that ranged from two hundred to four hundred feet in thickness. This formed a natural dam that trapped the river to form Earthquake Lake, a body of water some six miles long and as much as 180 feet deep. Much less spectacular, but far more costly, are the mudslides that have plagued parts of California within recent years. In some parts of the state, expensive homes have been carried away and destroyed by the blanket of unstable soil on which they were built. Other houses have become filled with sheets of syrupy mud that have flowed into the canyons where they were located. The mudslides, like rockslides, are the result of earth materials responding to gravitational attraction.

Not only do Earth's internal and external forces play important roles in shaping the landscape, they are intricately related to the processes that make rocks. The relationships of rock materials to geologic processes form a *rock cycle* that permits us to trace the various paths that rock materials follow and the processes that affect them along the way (FIGURE 17). This geological cycle was first noted in the late eighteenth century by James Hutton, who said: "We find no sign of a beginning—no prospect of an end."

How does the rock cycle work? Let us begin with a mass

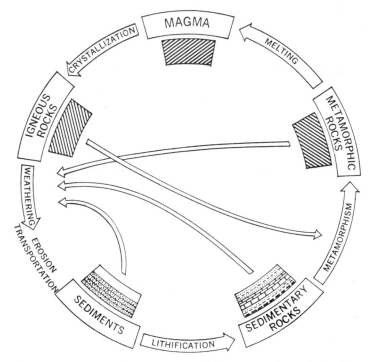

FIGURE 17 Reproduced by permission from *Physical Geology* (third edition) by L. D. Leet and S. Judson, Prentice-Hall, Inc. (1965). The rock cycle, shown diagrammatically. If uninterrupted, the cycle will continue completely around the outer margin of the diagram from magma through igneous rocks, sediments, sedimentary rocks, metamorphic rocks, and back again to magma. The cycle may be interrupted, however, at various points along its course and follow the path of one of the arrows crossing through the interior of the diagram.

of molten rock associated with volcanism and let it solidify to form an igneous rock. Upon exposure to the atmosphere and hydrosphere, the igneous rock will be weathered and eroded to produce sediments that may be carried away and deposited elsewhere. In due time these sediments may be covered by other rock materials and eventually become compacted and cemented together to form sedimentary rock. Now let us subject this deeply buried sedimentary rock—along with the remaining igneous rock from which it was derived—to intense heat, pressure, and the deforming forces of great earth movements. The igneous and sedimentary rocks cannot withstand these severe physical conditions and are likely to become converted into metamorphic rocks. If these metamorphic rocks continue to undergo increased heat and pressure, they may eventually melt to form magma. When this magma cools and solidifies, an igneous rock is formed and the cycle has been completed.

The rock cycle, like the hydrologic cycle, represents a response of earth materials to various forms of energy. The outer circle of Figure 17 represents the complete, uninterrupted cycle. But the cycle is not always completed, for there can be many "short circuits" along the way. Some of the shortcuts that bypass parts of the cycle are indicated by the arrows within the circle. Notice that the starting point in the rock cycle is magma. This molten rock is a key link in an endless chain that connects present-day rocks with the igneous rocks of the earth's first crust.

And so it has been and will be with this ancient planet. The processes of uplift, weathering, rock forming, and metamorphism have been repeated countless times during the 4½-billion-year history of its crust, and these same processes are still operating today and will continue to operate in the future. They remind us—sometimes quite forcefully—that Earth is not the passive, unchanging planet that it may seem. Far from it—this dynamic globe of air, land, and water is forever being modified by the many forces that are constantly at work both within it and on it.

7

IN SEARCH OF THE PAST

Man has speculated about the past since time immemorial, for the records of almost every civilization reveal some evidence of this search for ancient history beyond human recollection, written history, and the most ancient legend. Nor is this unusual, for man is a most inquisitive creature, and the scientist is an especially curious breed of man. This is particularly true of the geological scientist, for the earth abounds with myriad questions that literally cry out for answers. Indeed, Earth not only invites the geologist to study her features, but challenges him to explain their composition and origin.

Hopefully, the preceding chapters have revealed something of the nature of earth materials and how these materials may be changed by natural forces over a given period of time. But thus far in our endeavor to understand the earth, we have only considered the physical side of geology. Attempts have been made to answer questions such as what has happened to the earth and its materials as they have been modified by geologic processes, and what are the forces that have caused these changes? But the geologist is not satisfied merely to know *what* has happened; he also wants to know *why* it happened. Having learned that volcanic eruptions originate from concentrations of magma that lie deep inside the crust, the geologist next wants to know why these pockets of hot molten rocks exist and how they obtain their heat. In other words, when we learn the *effect* of a certain geologic phenomenon we then want to track down its *cause*. Although it is not always obvious to the observer, the relation between cause and effect is important in studying the earth. This is not to say that studies of cause-and-effect relationships will solve all geologic problems—far from it. Nevertheless, given the effect—which he can often observe at first hand—the earth scientist does have an intelligent point of departure from which to try to track down the causes of certain geologic phenomena.

Scientists of all disciplines typically base their investigations upon the "whats" and "whys" of their particular field of study. But the geologist must consider yet another parameter in his research: the factor of *time*. Thus, once the geologist has learned what happened and—hopefully— why, he then asks: *"When did it happen?"* The concept of "when" requires placing cause-and-effect relationships in a time continuum whereby geologic events can be put in their proper geologic sequence. It is not enough, then, for the geologist to know that volcanic eruptions are fed with magma from reservoirs of molten rock within the earth, or that it is quite likely that these rocks were liquefied by intense heat produced by radioactive elements in the crust. He looks at ancient lava flows and volcanic cones—the effects of volcanism—and wants to know at what points in geologic time particular eruptions took place.

The Record in the Rocks

The rocky layers of the earth's crust are like stony pages in a mammoth, earthen history book: an ancient "volume" that tells a fascinating story of natural forces that have been operating on Earth during its 4½-billion-year history. Almost every rock that the geologist studies will yield two important pieces of information. First, it will normally provide some evidence as to the conditions under which it was originally formed. Second, rocks frequently record some of the things that have happened to them *since* they were formed. Thus, the igneous rocks record the more fiery chapters of earth history: They tell of lava floods, tremendous explosions, and showers of glowing cinders and ash. Equally violent are the episodes represented by metamorphic rocks. These "made over" rocks reveal evidence of great crustal unrest and deformation, of a buckling crust, minerals that were compressed and heated, and new rocks formed from old.

Yet for all the information gleaned from the study of igneous and metamorphic rocks, it is the sedimentary rocks that provide our most valuable clues to earth history. More than any others, these rocks reveal the fascinating sequence of events that have helped shape the modern landscape. They contain bits of evidence that have recorded the comings and goings of vast inland seas, of powerful glaciers of rasping ice, of eroded lands and raging rivers, and of restless winds and drifting sands. Here, too, can be found most of the earth's fossils: the tracks, trails, teeth, and bones that reveal the parade of life throughout geologic time.

How does the earth historian "read" the record in the rocks? In one sense he is somewhat like the scholar who delves into ancient historic records in an attempt to work out the written history of mankind. But the techniques of the historical geologist are more closely related to those of the archaeologist, who studies man's buildings and artifacts to reconstruct the much more ancient chapters in the development of man. Like the history scholar and the archaeologist, the earth historian gathers clues and bits of evidence from a multitude of sources. He combs the rocks for signs of the many changes in geography, climate, and life that have occurred in the geologic past. From the topography, or general configuration of the land surface, he may obtain evidence of the geologic forces that have shaped the landscapes of yesterday and today. And the presence of fossils may permit him to reconstruct the ancient environments in which prehistoric plants and animals lived and died. In short, the earth historian pieces together earth history on the basis of information gleaned from a study of the earth's rocks and fossils. These materials of the earth's crust are the primary documents of historical geology, and they reveal the intriguing history of this ancient planet.

The first decipherable chapter of earth history is found in the most ancient rocks of the earth's crust. Formed at the very dawn of geologic time, these rocks are typically found deeply buried beneath younger rocks that have been deposited on top of them. This part of the rock record may be deep within the earth, and it is for this reason that geologists work from the "bottom up" when reconstructing earth history. Although they have been defaced by time and subsequent geologic events, the earliest formed rock layers provide much information about the opening episodes in Earth's long history. The more recent chapters are found in the younger, upper rocks. But "reading" earth history is not as simple as reading a book. The geologic formations in many parts of the world have been greatly disturbed and are not always found in the sequence in which they were formed. Elsewhere, the record in the rocks has been completely obliterated by erosion or greatly altered by metamorphism. Faulting and other structural disturbances have also taken their toll, for some of Earth's rocky "pages" have become shuffled and out of sequence, and many have been lost forever. These missing pages in the geologic record make the reconstruction of earth history a more formidable task, but the geologist has learned not to be discouraged. He has found that some of the gaps in the record may have been preserved in other areas and that he is frequently able to "fill in the blanks" with this information.

Fossils: Silent Witnesses of the Past

In 1799, a young surveyor named William Smith was asked to plan the route of a new canal to be excavated in southern England. This assignment called for some geologic knowledge of this area, for Smith knew that the type of rock to be encountered would directly affect the cost of building the canal. There were no engineering geologists in those days, so the young surveyor had to conduct his own geological studies. This could have been a serious handicap, but as luck would have it, Smith's hobby was collecting rocks and fossils—an avocation that he soon put to practical use in the field.

As his workmen dug through the different layers of rocks, Smith found that the various rock strata could be identified by the kinds of fossils that they contained. More important, he soon learned that the fossils found in each rock layer differed from those in the rocks above and below it. Smith quickly recognized that the fossils were useful field guides, and he soon devised a technique whereby he could predict the location and physical characteristics of buried rocks by means of the fossils he found exposed in rocks and quarries. Smith's technique of "matching" the rocks and fossils from different areas, known today as *correlation,* is one of the most important processes used in interpreting the record in the rocks. As a result of his work, "Strata" Smith—as he was later appropriately nicknamed—constructed the first geologic map of England, Wales, and a part of Scotland.

Although "Strata" Smith was one of the first to put fossils to some practical use, he was certainly not the first to notice these fossilized remains of prehistoric plants and animals. There is much evidence to suggest that traces of ancient plant and animal life began to interest man at a very early time, for they have been found in association with the remains of certain primitive and prehistoric men. These early humans apparently thought that fossil shells, bones, and teeth were "good medicine," or that they possessed the ability to bring good fortune or remove curses. Interestingly enough, this idea still prevails in some more primitive cultures today, for tribal medicine men may include fossils among the many items found in their healing kits.

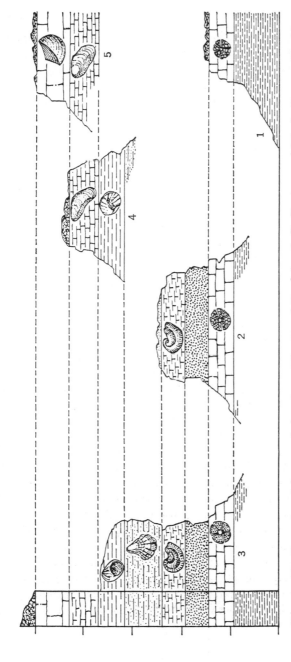

FIGURE 18 Reproduced by permission from *The Fossil Book* by C. R. Fenton and M. A. Fenton, Doubleday & Company, Inc. (1958). How fossils are used to match, or correlate, beds of similar ages. The complete section of these varied strata is shown here.

At a much later date, fossils attracted the attention of certain of the early Greek and Roman philosophers. As early as the sixth century B.C., Anaximander of Miletus reported fossil fish high above sea level and concluded that they were the forerunners of all living forms. Xenophones of Colophon (576–480 B.C.) was another observant Greek who noticed the fossilized remains of shells, fish, and other marine organisms embedded in rocks far from the ocean. Xenophones correctly inferred that these objects were the remains of creatures that inhabited seas that had formerly covered this part of the earth. And about 450 B.C., Herodotus reported the presence of fossils in Egyptian limestones. Because of their shape, he assumed that the remains of these one-celled organisms were petrified lentils left over from the food supplies of the slaves who erected the pyramids.

The venerable Greek naturalist Aristotle was also interested in fossils. But although he did recognize fish remains as fossils, he thought that they were the work of supernatural agents, such as mysterious "plastic forces." Theophrastus, one of Aristotle's students, speculated that the fish had wandered in from the sea by means of subterranean rivers, or that the fish eggs had been left behind in the earth to hatch at some later time. As noted in Chapter 2, the development of paleontology was impeded during the so-called "Dark Ages." During this time it was heresy to consider fossils as anything but the remains of attempts at special creation, freaks of nature, or devices of the devil placed in the rocks to lead men astray. These superstitious beliefs hindered the scientific study of fossils for hundreds of years.

Within the past century, however, fossils have generally been accepted without question as being the remains of ancient life. They have become a valuable tool of the geologist, and paleontology has done much to advance our knowledge of prehistoric life. How long has life been present on Earth? No one knows for sure, but the oldest known fossil remains—primitive bacterialike organisms—

are estimated to be more than three billion years old. Yet despite the great antiquity of life, the first clear and unbroken record of organic activity began some six hundred million years ago and can be traced through its evolution into the more advanced forms of later geologic time and today. The chain of past life begins with the remains of rather primitive and relatively simple organisms found in very ancient rocks. However, comparison of these organisms with fossils of similar species that lived in later time shows that fossils become progressively complex and more advanced in the younger rocks. As might be suspected, this increasingly complex life continuum has lent considerable support to the theory of organic evolution.

Concerned as it is with the record of life, paleontology is closely related to the biological sciences and utilizes many biological concepts and techniques. The student of fossils also relies heavily on the Principle of Uniformitarianism, for he must assume that the present provides clues to the past. Because of the vast spans of time involved in earth history, we cannot always be certain of the life habits or environments of extinct plants and animals. However, when we discover a fossil group whose members closely resemble the members of a living group, we can usually infer that the fossil organisms lived under conditions similar to those of their living counterparts. Yet even if the fossil form has no living counterparts, all is not lost, for most fossils furnish some clue as to where, how, and when they lived. They may even provide us with a record of how they died.

Why is it helpful to know where and when a prehistoric organism lived? One reason is that fossils help us reconstruct ancient environments. These, in turn, can be used to work out the changing geographic patterns for various times in earth history. It is known, for example, that most varieties of plants live on the land. The presence of fossil trees or stumps *in situ* (in place where they originally grew) suggests a land environment, as would the bones of land animals. Rocks that contain fossilized remains of such terrestrial organisms were probably formed on land. Elsewhere, the geologist may find fossils indicative of swamp or marsh conditions, or a mixture of leaves with fresh-water clam and snail shells that might reveal a lake or stream environment.

Other plants and animals have always apparently been restricted to life in the sea. The presence of marine fossils normally suggests that the rocks containing them probably formed from sediments that were laid down on the floor of an ancient sea. In some areas there may be an intermingling of fossils of marine and terrestrial origin. This mixing occurs when a fresh-water stream enters a body of salt water and is indicative of deltaic deposition. By plotting the distribution of marine strata in relation to rocks formed on land, the geologist can construct a *paleogeographic* map that will show the extent of prehistoric seas and the location of ancient continents. Maps of this type not only help reveal how the earth has changed at various times, they can also contain clues that may lead to the discovery of valuable mineral deposits.

The fossilized remains of elephants have been found well within the Arctic Circle, and fossil ferns have been found in Antarctica. Animals such as elephants could not, of course, live in these areas today, for they have become adapted to much warmer climates. What is the significance of their occurrence in the polar regions? Clearly, the climate in these areas was once much milder than it is today. By the same token, the presence of fossil reindeer in France and "Ice Age" musk ox in Arkansas and New York show much colder climates for these areas. Assuming, then, that most living organisms have always inhabited similar climatic regions, fossils can also be quite useful in reconstructing climates of the geologic past.

Fossils also have a practical value, for many important mineral resources are known to occur in fossiliferous rocks. Using "Strata" Smith's correlation technique, the economic geologist uses fossils as clues in the search for rocks that contain valuable deposits of ore, coal, oil, and natural gas. Micropaleontology (the study of fossils so small that they

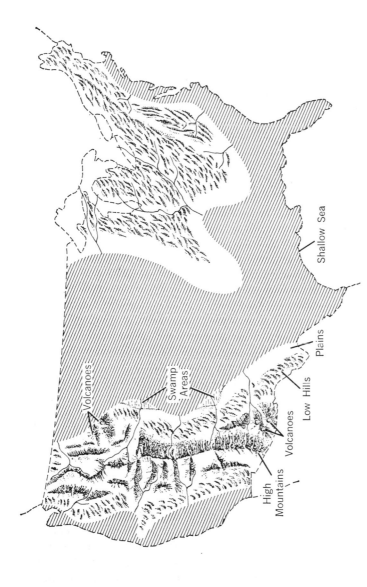

FIGURE 19 This generalized paleogeographic map shows the United States as it might have appeared during Cretaceous time, about eighty million years ago. (U. S. Geological Survey)

are best studied with a microscope) is particularly useful to the petroleum geologist. The tiny microfossils are so small that they are not likely to be broken by the drill bit as it probes the earth in the search for oil. Thus, they can be brought to the surface with little or no damage to their original structure. The micropaleontologist washes the rock cuttings that have been taken from the bore hole of the well and separates the microfossils from the surrounding rocks. The microfossils are then mounted on special slides and studied under the microscope. Information derived from these minute plant and animal remains may provide valuable data on the age of the subsurface geologic formation as well as its potential as an oil-producing horizon.

How Old Is Old?

Stonehenge, the Roman Forum, and Pompeii are among the most famous and best known of the world's historic attractions. Yet as different and widely separated as they are, all of these ruins have a single common denominator: great antiquity. And while these fascinating relics pose many mysteries, either directly or indirectly most of these problems revolve around the age of these archaic monuments. This is not surprising, for "How old?" is a question that is commonly asked about everything from infants to the moon. Curiously enough, most of us are interested in almost *anything* that is old. And, generally speaking, the older an object is, the more interest it generates. This is true whether we are dealing with coins, antique furniture, dinosaur bones, or the Earth.

Man's interest in time is, of course, quite natural. Ours is an age when time and its measurement are most important, for we live in a time-ordered culture in which our every activity is rather closely regulated by clock and calendar. Yet, as important as time is to most people, it is of far greater importance to those who search for the past. Without some knowledge of time, historic and prehistoric events cannot be placed in their proper sequence, and their value is thereby minimized. The history scholar is faced with this problem as he attempts to correlate historic records over many decades and centuries. The archaeologist must deal with still larger blocks of time, for he delves into cultures that existed several thousands of years ago. But the geologist is concerned with time spans that literally boggle the mind. Indeed, someone has said that the geologist can easily be distinguished in a crowd by the way he casually tosses around millions and billions of years!

There is good reason for this. The objects of geology—rocks, minerals, and fossils—are by their very nature quite ancient. And without some knowledge of their relative age, the geologist cannot place them in their proper sequence in earth history. Why their *relative* age? We commonly mark the passage of time by relating it to a series of events. It may be a certain season of the year, a particularly severe storm, or a general period of life such as "youth." The geologist also uses a relative system of dating whereby he can state that one geologic event—say a volcanic eruption —occurred before or after another event such as a rise in the sea level. He does this in the field by comparing the relative positions of the rocks and by studying any fossils that might be in them. But although this may provide some indication of the age of one geologic event or rock unit in relation to another, it does not imply age in years. Rather, it is somewhat like learning that the American Revolution preceded the Civil War. Although no *absolute* time (time measured in years) is given, it does place these events in their proper historic perspective.

The proper sequence of a series of historic events can normally be established by various written documents and calendars. But in determining a relative sequence of geologic events, the rocks are our documents and the Principle of Superposition is our sequential guide. As noted earlier, this basic principle assumes that in an undisturbed sequence of stratified rocks, the topmost layers are youngest and the beds beneath them get progressively older as we go down into the crust. This is true, of course, because this is

the order in which they were originally formed. Moreover, the fossils contained in each successive rock unit will
be recognizably different from those in the formations that
lie above and beneath them. This is known as the Principle
of Faunal Succession.

Earth historians, like historians who deal with the
history of civilization, must have some means of commu-
nicating these relative times to each other. Using data derived from superposition and faunal succession, geologists
have devised a special geologic time scale composed of
named units of time during which the various rock units
were formed. The largest of these time units are called
eras; each era is divided into *periods,* which in turn may
be divided into still smaller units known as *epochs.* Ar-
ranged in chronological order, these time units form a sort
of geologic "calendar" that provides a standard by which
the age of the rocks can be discussed. Unlike days and
years, the units of the time scale are arbitrary and of unequal duration, for there is no way of knowing the exact
amount of time involved in each interval. The time scale
does, however, make it possible to put a "time tag" on
the various chapters of earth history. Thus, when the geologist says that a fossil is Mesozoic in age, he is referring
to the remains of a plant or animal that lived during the
Mesozoic Era, about sixty-five million years ago. Each
of the time values on the scale provides some indication as
to when the organism lived but in terms of relative time
rather than absolute time.

The need to subdivide earth history into manageable
"chapters" was recognized rather early in the development
of geology. One of the first attempts at subdivision was
made in 1756 by Johann Lehmann, who developed a
relative time scale on the basis of rocks exposed in central
Europe. As a result of extensive field work in Germany,
Lehmann recognized three classes of mountains and three
classes of rocks that formed them. This pioneer work was
followed by a fourfold classification of rock and time proposed by Abraham Gottlob Werner, whose Neptunist

GEOLOGIC TIME SCALE

ERA	PERIOD	EPOCH	SUCCESSION OF LIFE
CENOZOIC RECENT LIFE	QUATERNARY 0-1 MILLION YEARS	Recent / Pleistocene	
CENOZOIC RECENT LIFE	TERTIARY 62 MILLION YEARS	Pliocene / Miocene / Oligocene / Eocene / Paleocene	
MESOZOIC MIDDLE LIFE	CRETACEOUS 72 MILLION YEARS		
MESOZOIC MIDDLE LIFE	JURASSIC 46 MILLION YEARS		
MESOZOIC MIDDLE LIFE	TRIASSIC 49 MILLION YEARS		
PALEOZOIC ANCIENT LIFE	PERMIAN 50 MILLION YEARS		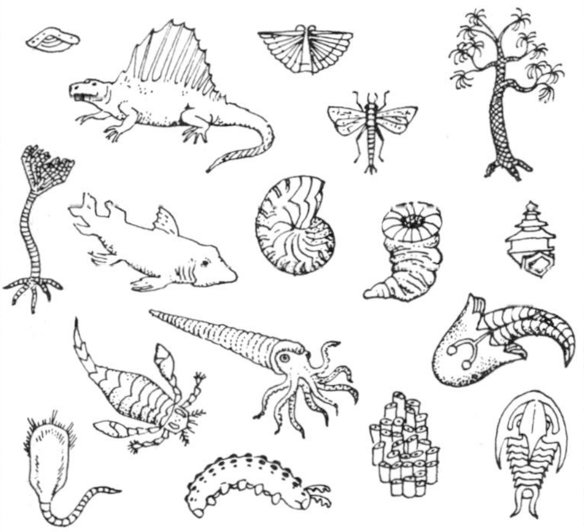
PALEOZOIC ANCIENT LIFE	CARBONIFEROUS — PENNSYLVANIAN 30 MILLION YEARS		
PALEOZOIC ANCIENT LIFE	CARBONIFEROUS — MISSISSIPPIAN 35 MILLION YEARS		
PALEOZOIC ANCIENT LIFE	DEVONIAN 60 MILLION YEARS		
PALEOZOIC ANCIENT LIFE	SILURIAN 20 MILLION YEARS		
PALEOZOIC ANCIENT LIFE	ORDOVICIAN 75 MILLION YEARS		
PALEOZOIC ANCIENT LIFE	CAMBRIAN 100 MILLION YEARS		
PRECAMBRIAN ERAS			
PROTEROZOIC ERA			
ARCHEOZOIC ERA			

APPROXIMATE AGE OF THE EARTH MORE THAN 4 BILLION 550 MILLION YEARS

FIGURE 20 Reproduced by permission from *Fossils: An Introduction to Prehistoric Life* by William H. Matthews III, Barnes & Noble, Inc. (1962).

140178

Theory has been previously discussed. This eighteenth-century mineralogist believed that all rocks had been precipitated from sea water and could be assigned to one of four major subdivisions. Although Werner's theory of rock formation did not hold up, his idea of establishing distinctive time units persisted. With the growth of geologic thought during the nineteenth century, geologists in western Europe and the British Isles became increasingly aware of the need of some scheme to develop a geologic time scale. It is due to their efforts that we have a geologic column and time scale that is in universal use today.

One of the major problems in devising the geologic time scale was the establishment of criteria to separate the major episodes in earth history from one another. These natural breaks in geologic history were eventually based on supposedly worldwide geologic events of such magnitude as to be clearly discernible in the rock record. Thus, intervals of mountain-building activity and times when the seas changed their positions were generally recognized as events of sufficient importance to establish the beginning and end of the eras and periods of the geologic time scale. We now know that mountain-building movements may be restricted to a single continent and are not necessarily universal. Nor did the seas go "in and out" with determinable regularity. Consequently, correlation of the succession of fossil assemblages in the rocks and the basic tenets of superposition are the bases for the standard rock column and geologic time scale as we use it today.

Numbering the "Pages" of Time

Despite the convenience and widespread use of the geologic time scale, it serves only to relate one geologic event to another. Thus, geologists soon recognized the need for a more quantitative "calendar"—one that would provide more absolute values in terms of actual years. In other words, some method was needed to date or "number" the rocky "pages" that have recorded earth history. But how

was such a table to be devised? How could the age of the
earth be translated into years?

This was not a simple task, and early scientific attempts
to determine the length of geologic time were understand-
ably well off target. However, it was no fault of the
scientists, for they had a minimum of critical information
upon which to build. One of the earliest attempts to date
the earth was made in 1715 by Edmund Halley. This
famous English astronomer—for whom Halley's Comet is
named—realized that the oceans began as a body of fresh
water and had become increasingly saline with the passage
of time. Why had the sea become so salty? In Halley's own
words: ". . . the saline particles brought in by the rivers
remain behind, while the fresh evaporate." Thus, even
in that early day, Halley and others knew that salts had
been dissolved from the rocks and carried to the sea by
streams. Halley suggested, moreover, that the total amount
of salt in the sea might provide some measure of the age
of the oceans, and hence give an indication of the age of
the earth. At the time that Halley proposed his salinity
method, the data necessary for computing the sea's age
were not at hand. But in 1898, the Irish scientist John
Joly felt that he had collected sufficient information to
make a reasonable estimate of the age of the oceans. He
calculated that it had taken eighty to ninety million years
for the sea to reach its present degree of salinity, and so
the earth was bound to be somewhat older than that.

At about the same time, other geologists tried to determine how long it has taken to form all of the rock layers
in the earth's crust. As a start, geologists conducted a series
of experiments to measure the rates of accumulation of
various sediments. They wanted to know, for example, the
length of time required to deposit the sediment needed to
form one hundred feet of limestone or one thousand feet
of sandstone. An attempt was then made to assemble data
on rock outcrops all over the world in order to determine
the maximum thickness of rock formed during each period
of geologic time. By adding together the thicknesses of all

of these beds, it was hoped that this total would give a reasonable approximation of the earth's age. It is obvious that there are many loopholes in such a method, and these early geochronologists were not unaware of them. They realized at the outset that different sediments accumulate at various rates in response to changes in the depositional environment. More important, however, there was no way of accounting for the amount of time represented by the erosional gaps scattered throughout the geological record. These unconformities—some of which encompass tens of millions of years—represent missing pages in earth history that are still bothersome today. Despite these problems, this method was used to provide age estimates that ranged from slightly less than one hundred million to more than four hundred million years old.

Another dating technique was proposed at the end of the nineteenth century by Lord Kelvin of England. Kelvin, a pioneer British physicist, based his dating method on the assumption that the earth was originally molten and has reached its present solid state by slowly cooling throughout geologic time. By estimating the amount of heat escaping from the earth, Kelvin's calculations indicated that the earth had solidified between twenty million and forty million years ago.

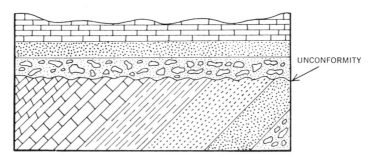

FIGURE 21   Unconformities represent gaps in the geological record. (U. S. Geological Survey)

Although considerably more accurate than the six-thousand-year date of Creation estimated by some biblical scholars, these early dating methods still fell quite wide of the mark. And so geochronologists kept searching for some means of transforming their relative time scale into an absolute time scale, whereby geologic events could be dated in terms of years. Then—as is so often the case in scientific breakthroughs—a chance discovery suddenly provided the key to the puzzle of geologic time. It happened in 1896 when Henri Becquerel, a French physicist, accidentally learned that certain natural substances such as uranium give off energy. This unexpected discovery attracted the attention of other physicists, who were fascinated by this newly discovered phenomenon and wanted to know more about it. Through the work of such dedicated scientists as Ernest Rutherford, Marie and Pierre Curie, Frederick Soddy, and others, it was eventually learned that certain rocks possess the property of radioactivity. In brief, radioactivity can be defined as the spontaneous disintegration of some elements from an unstable condition to a more stable condition.

The discovery of radioactivity stimulated additional research on radioactive substances and resulted in knowledge about rates of decay, the products of decay, and the types of energy given off during radioactive decay. Another significant stride was made in 1907 when Bertram Boltwood, American physicist and chemist, found that as uranium undergoes radioactive disintegration, lead is formed as a final product of decay. This discovery led him to suggest that natural radioactivity might provide the long-sought geologic "clock" that would provide a means for measuring the age of rocks. How? Boltwood assumed that if he could measure the amount of the uranium parent material remaining in a particular mineral and compare this with the lead produced by radioactive decay (the daughter product), he could then determine when the mineral was formed. He reasoned further that this would also provide the rate at which the parent mineral had broken down. Fortunately, Boltwood was correct on both counts, for both of these measurements can be made.

To understand the mechanism of this radioactive chronometer, let us briefly review the basic principles behind it. First, not all rocks contain radioactive minerals; thus radiometric dating is limited to those rocks that do. However, when certain masses of molten rock cooled and solidified in the geologic past, radioactive elements did become incorporated into certain mineral crystals. These radioactive minerals—for example, uranium and thorium —have large unstable atoms that undergo slow, spontaneous disintegration. The rate at which radioactive decay proceeds is constant—it is not affected by changes in heat, pressure, or chemical conditions. As the parent element disintegrates, new materials will be formed, and these can clearly be recognized and measured. For example, when uranium decays, helium (a gas) is released, and a new series of elements is formed. The last substance formed in this series is a form of radioactive lead. By computing the ratio between the radioactive lead and the remaining amount of uranium present in a given specimen, it is then possible to determine how long ago the radioactive mineral was formed.

The rate of decay of radioactive elements is expressed in terms of its half-life: the time required for half of the atomic nuclei in a sample of that element to decay. Various radioactive substances have half-lives that range from a fraction of a second to billions of years. But, with the exception of carbon 14, the radioactive materials used to date rocks have very long half-lives. One of the most widely used radiometric dating methods utilizes uranium 238 (U^{238}), which disintegrates through a distinct series of fourteen steps resulting in the decay product of lead 206 (Pb^{206}). The time it takes for half of the original amount of material to change to Pb^{206} is a constant 4.51 billion years. The incredibly long half-life of U^{238} renders it especially useful in dating extremely old rocks. In fact, it is this method that was used to date the oldest rocks yet discovered. These ancient intrusive igneous rocks are located in Canada and Africa and have been assigned ages of about 3.3 billion years. Other radioactive elements that can be used include rubidium (Rb^{87}), which has a half-life of 47 billion years and decays to form strontium 87 (Sr^{87}) and thorium 232 (Th^{232}), whose daughter element is lead 208 (Pb^{208}) and which has a half-life of 13.9 billion years. Unfortunately, these radioactive elements are not much more abundant than uranium, and consequently are of limited use. Another dating method involves the use of potassium 40 (K^{40}), a rather common and widespread form of potassium that exhibits feeble radioactivity. The disintegration of K^{40} is very complex, but one of its daughter products is a form of argon (A^{40}) that can be detected in minute quantities. The fact that potassium is a common constituent of igneous rocks plus, the K^{40} half-life of 1.31 billion years, has made the potassium-argon method the most widely employed radiometric dating technique now in use.

The radioactive elements described above have very long half-lives, and "atomic clocks" that are energized by these isotopes "run down" very slowly. Consequently, the amount of daughter elements formed within a few million years is so small that these decay products cannot be accurately measured. Luckily, the problem of how to date much younger objects was solved with the development of radiocarbon dating. Especially useful in dating material less than fifty thousand years of age, this technique was developed in 1949 by Willard F. Libby, who was awarded the Nobel Prize in chemistry for this discovery. Libby found that nitrogen (N^{14}) in the outer atmosphere is continually bombarded by cosmic rays; this creates C^{14}. The radioactive carbon thus produced unites with oxygen to form carbon dioxide which—along with the C^{14}—is obtained from the atmosphere by plants. Living animals pick up C^{14} from plants or from the plant-eating animals that they consume. As long as the plant or animal lives, a balance is maintained between the radioactive carbon and the ordinary variety of carbon, for new C^{14} is added as fast as the old disappears. But when an organism dies, it no longer takes in C^{14}, and the radiocarbon in its remains is gradually lost. The decay of this radioactive carbon proceeds at a known rate, and this rate is such that one-half of the C^{14} has disintegrated at the end of 5568 years. In using this technique, the quantity of C^{14} remaining in the sample must first be determined. Then the approximate age of the specimen is ascertained by comparing the ratio of radioactive carbon remaining in the specimen to the amount assumed to be present in most living things. Thus, if the ratio is one-half that in a living organism, this indicates that the plant or animal lived about 5568 years ago. If it is one-fourth, the organism lived some 11,136 years ago. This method has been used to date a variety of relatively recent specimens including wood, cloth, bone, shells, flesh, and hair.

Historians, archaeologists, and anthropologists have found this radiocarbon dating to be especially valuable. They have learned that this technique is quite dependable, for they have also checked this method against items such as human artifacts and tree rings whose ages have been determined by other means. These dates are in close agreement, and this indicates a high degree of accuracy for radiocarbon dating. Some objects whose ages have been determined by the use of C^{14} include linen from the Dead Sea scrolls in Israel, 1983 years; a loaf of bread from the ruins of Pompeii, 1995 years; and charred wood from the hearths of Neanderthal men of the Middle East, about thirty thousand years.

Radiometric dating has been a great boon to the geologist, for it has helped to calibrate the geologic time scale into more absolute values and provides a means whereby geologic events can be placed in their proper chronological sequence. By means of one or another of these radioactive chronometers, age determinations have been made for thousands of rock masses and human artifacts. In many instances these objects have been crossdated by two or more methods as a check against their validity. By and large, the results of these checks indicate a reasonable degree of accuracy for our radioactive time clock. What are some of the more significant findings of the geochronologist? A composite of thousands of analyses has revealed the following highlights in Earth history. The earth appears to have solidified about 4.5 billion years ago, and the first extensive rock masses were formed about one billion years later. Although the oldest known life forms appeared more than three billion years ago, the most dependable fossil record starts with the dawn of Cambrian time (FIGURE 20), some six hundred million years ago. The beginning of the Pleistocene Epoch—during which man appeared on earth—is placed at about two million years ago.

To appreciate more fully the enormity of geologic time and to place the above events in a more familiar frame of reference, consider the following:

If we imagine the whole of earth's history compressed into a single year, then on this scale, the first eight months would be completely without life. The following two months would be devoted to the most primitive creatures ranging from viruses and single-celled bacteria to jellyfish, while mammals would not have appeared until the second week in December. Man, as we know him, would have strutted onto the stage at about 11:45 P.M. on December 31. The age of written history would have occupied little more than the last sixty seconds on the clock.*

* Richard Carrington, *The Story of Our Earth* (Harper & Brothers, 1956), pp. 47–48. Reprinted by permission of the publisher.

8

FITTING THE PIECES TOGETHER:

THE GEOLOGIST AT WORK

Off the coast of California, three divers hover above the bottom of the sea. They may look like "scuba" divers enjoying their hobby, but they are actually marine geologists mapping and sampling the ocean floor. Later these maps and samples may yield clues to help unlock the ocean's treasure chest of mineral wealth.

Across the continent, at Massachusetts Institute of Technology, a geophysicist feeds data into a computer. It is programmed to generate five million models of the earth. Hopefully, this computer will give information that will help construct a reasonable theoretical or mental model of Earth's interior structure.

Geologists in the crater of a Hawaiian volcano carefully probe still-smoking ash with special glass tubes. They are collecting volcanic gases that will later be analyzed in the laboratory. Data derived from these analyses may provide information that will help explain the volcano's behavior.

In Flagstaff, Arizona, a scientist carefully studies the moon through a thirty-inch telescope. An astronomer? No. He is an astrogeologist, whose telescopic observations and photographs will be used to interpret lunar geology and compile maps so vital to moon exploration.

A solitary unmanned satellite continues its methodical orbit around Earth while sensitive instruments steadily record a wealth of scientific information. Earthbound geologists in Washington carefully monitor this orbiting laboratory's progress and eagerly await each of its messages. They are especially interested in the infrared images that point out "hot spots" in the earth's crust. These places —where molten rock is relatively near the surface—may bear watching: They could mark the sites of future volcanic activity. They are equally interested in the magnificent photographs sent back from space, for these greatly facilitate the important task of mapping Earth's surface. Meanwhile, a group of mining geologists in Nevada concentrates on a radar image that has been beamed earthward. They are pleasantly surprised to learn that radar imagery has disclosed geologic structures that had not been revealed by earlier field mapping or photographs taken from aircraft. There is good reason to believe that these newly discovered structures may contain valuable deposits of gold.

The above vignettes clearly indicate the varied ways in which geoscientists study the earth, and they suggest, in small part, the many tools that the geoscientists use. This is, of course, a far cry from the days of Werner and Hutton, for these geological pioneers did not have access to the fact-finding techniques used by modern earth scientists. Be that as it may, modern geologists owe much to these early students of the earth, for they laid the foundation for much of geology as we know it today.

How do today's geoscientists go about the tasks of locating and fitting together the pieces in the great "puzzle" called Earth? And how do they use this mass of data to answer questions about this fascinating—but often perplexing—planet? The geological scientist obtains his data from every imaginable source of information—direct observation in the field, data from airplanes and spacecraft, models developed by computers, and exhaustive laboratory and statistical studies.

But despite the type of problem that is involved or how the data have been gathered, all geoscientists use the problem-solving technique known as the *scientific method*. This is not some mysterious thought process available only to the scientist in his "ivory tower." On the contrary, the scientific method is a straightforward, systematic, investigative approach that can be applied to *any* problem or question. How does it work? It begins with the asking of a question: for example, how, when, or why did a certain geologic event occur? Next comes the collection of evidence or data that can be used to make predictions or arrive at preliminary conclusions. These conclusions are usually presented in a number of hypotheses that can be supported by the available data. Multiple hypotheses are better than a single hypothesis, for they make possible a wider interpretation of the scientific facts. Once reasonable hypotheses have been formulated, they must then be tested to determine which one of them best fits both the observed and deduced facts. More often than not, this requires further predictions and modifications of some hypotheses and the outright rejection of others. Finally, it is necessary to decide which one hypothesis most closely fits the data and most logically answers the question at hand. This system is workable in geology because the laws of Nature express regular or systematic relationships. Thus, the unknown can commonly be predicted from what is known.

How does one test the hypotheses that have been developed by the scientific method? Unfortunately, means for testing and proving geological hypotheses are not always at hand. When it is possible to investigate geologic phenomena directly, the geoscientist uses many of the experimental techniques of the chemist, physicist, and biologist. But at other times he must operate like the archaeologist, for his evidence is indirect, and clues are widely scattered throughout the geologic record. Worse yet, they are commonly obscured by the ravages of time. More often than not, the geologist is denied the direct experimental approach to the study of geologic phenomena. How, for example, does one study the earth's interior or observe at first hand the causes of continental glaciation? These are not possible, of course; hence the geologist must base his solutions to these and similar geologic problems on a large number of varied and separate pieces of information. Thus, some aspects of geologic problem solving are classic examples of inductive reasoning.

Within recent years geologists have increasingly relied on mental models to explain certain phenomena that could not be studied at first hand. Consider, for example, the interior of the earth. No one has ever seen the inner regions of our planet, nor is it likely that this will ever be possible. Yet in Chapter 5, the interior of the earth is described in some detail. How do we know there is a crust, mantle, and an inner and outer core? And how can their relative thicknesses and compositions be represented? This "picture" is presented by means of a mental or theoretical model that translates into visual form an idea that exists about a particular geologic object or process. Thus, the standard model of the earth's interior has been developed by geoscientists over many decades of careful research.

Thanks to our modern age of computerization, geoscientists are now constructing mathematical models and taking a more quantitative approach to the study of certain geologic processes and objects. They have found that some verbal descriptions can be reduced to numerical measurements, and by the use of mathematics they can express the relationships among the numbers. These permit the geologist to construct mathematical models to explain geologic phenomena that cannot effectively be illustrated by other means. Using this technique, geologists in the Soviet Union and the United States are generating computer models that have yielded much valuable information about Earth's interior. They have found that the inner earth is considerably more complex than was originally suspected and that estimates of dimensions of the core and mantle may be off by miles.

But not all geologic mysteries lend themselves to quantitative study. Many problems are so complex and the geologic record is so incomplete that mathematical analyses and computer programs cannot be applied to them. Thus, the sound reasoning of the scientific method is still very much in use, even in this booming age of computerized quantification.

You will recall that geology is an eclectic science in that it draws heavily on its sister sciences. Because geologists collect such varied types of data, they should ideally be expert mathematicians, physicists, chemists, and biologists before ever undertaking a study of the earth. Although expertise in each of these areas is certainly not always possible, the geologist must, nonetheless, have a broad fundamental knowledge of many scientific fields. Still more important, he must know when to call on the help of specialists in other fields should the need arise. In short, the geoscientist gathers his information from multitudinous sources and relies heavily on the techniques and concepts of all the basic sciences. To understand better the way geologists think and the methods by which they gather, interpret, and correlate a wealth of diverse data, let us trace the development of a scientific idea from its birth through its evolution to a full-blown theory. Let us follow the development of the Theory of Continental Drift, a controversial idea that has been debated by earth scientists for decades. This is an "on-again, off-again" type of theory that has captured the interest of scientists from many different fields.

As early as 1620, Sir Francis Bacon, the famous English

philosopher, was intrigued by the remarkable similarity of the Atlantic coastlines of Africa and South America. More than two hundred years later, Antonio Snider noted the resemblance between certain fossil plants that had been collected in both Europe and America. And, like Francis Bacon, Snider also recognized that the coasts of Europe and America matched each other rather closely. In his attempt to determine the geographic distribution of these ancient plants, Snider fit the continents together like matching pieces in a giant jigsaw puzzle and theorized that they had formerly been part of one master landmass. By 1885, an Austrian geologist named Edward Suess advanced the idea that the geological formations in the Southern Hemisphere were so similar that he could fit them together in a single, massive continent that he named Gondwanaland.

But it remained for Alfred Wegener, a German meteorologist, to tie these dangling ends together. This he did in 1915, when he proposed his now-famous Theory of Continental Drift. Wegener's proposal was most controversial, for it held that all of the world's landmasses had once been joined together in one great supercontinent that he named Pangea, a word which literally means "all earth." According to Wegener, this mammoth protocontinent broke apart some two hundred million years ago and the pieces slowly drifted to their present locations. Although this idea is still supported by some, more recent evidence favors the concept of two original landmasses: Laurasia in the Northern Hemisphere and Gondwanaland in the Southern. Supporters of the two-continent theory hold that Laurasia consisted of Eurasia and North America, while Gondwanaland was composed of Australia, Antarctica, the Indian subcontinent, Africa, South America, Malagasy, and various submerged fragments (FIGURE 22).

To support his hypothesis, Wegener pointed to the previously noted jigsaw puzzle fit of the coastlines on either side of the Atlantic. He also reconstructed the climate and life forms of the world at certain points in earth history. The evidence Wegener accumulated strongly suggested

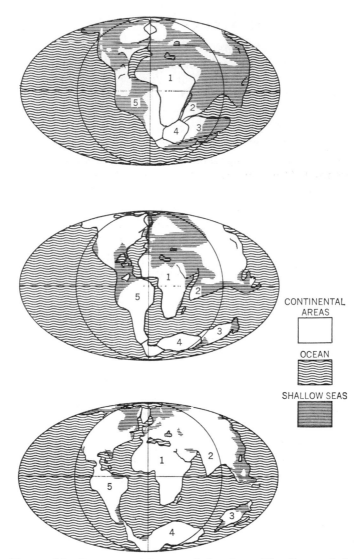

CONTINENTAL AREAS

OCEAN

SHALLOW SEAS

FIGURE 22 Reproduced by permission from *The Crust of the Earth* by Keith Clayton, The Natural History Press (1967). The maps illustrate three stages of Wegener's reconstruction of the breakup of Gondwanaland, the old supercontinent of the Southern Hemisphere. Area marked (1) is Africa, (2) India, (3) Australia, (4) Antarctica, (5) South America.

that during certain intervals of geologic time the rocks, climates, plants, and animals had been quite similar in different parts of the globe. Therefore, it seemed logical to assume that the now widely separated continents had been much closer together in the geologic past.

Wegener carefully documented all data that supported his theory, but his critics accused him of being less conscientious in considering data that conflicted with his unique proposal. More important, he was unable to provide satisfactory answers as to how, when, and why his supercontinent split apart and its pieces began to drift. Wegener's only suggestion was that since there was general agreement on vertical movements of the continents, why could there not also be horizontal movements of the landmasses? What force could be great enough to trigger these lateral movements? Wegener explained that the continents had moved laterally in response to the force of Earth's rotation or the gravitational attraction of the earth's equatorial bridge. But, although these forces do exist and do exert some force, their effect is not great enough to cause continental movements. And so as objections mounted, Wegener's idea of drifting continents was gradually cast aside and was either ignored or ridiculed by most earth scientists.

But today—after being in mothballs for years—continental drift is in vogue again. What brought about the rejuvenation of this all-but-extinct idea? The resurrection of Wegener's theory illustrates well the "conglomerate" nature of geology. It reveals the varied tools and techniques that the geoscientist applies in the solution of geological problems and is a classic example of how geologists use data derived from the other basic sciences to further our understanding of the earth. It shows, moreover, that the information provided by physics, biology, chemistry, and astronomy has been as "bread cast upon the waters." In other words, much geologically derived information has helped to solve several long-standing problems in the other basic scientific disciplines.

Although supporting evidence has come from diverse

—and often surprising—sources, geophysical research has probably played the most significant role in revitalizing the drift theory. This is especially true of recent investigations dealing with paleomagnetism, or the study of the earth's magnetic field as it existed in the geologic past. These same studies have lent support to another controversial hypothesis: the idea that the ocean floors are spreading as a result of convection currents (FIGURE 15) in the mantle. This is especially significant in the consideration of continental drift, for sea-floor spreading may be the mechanism needed to transport the continental masses.

Physicists have long realized that Earth's magnetic field behaves as if a bar magnet were in the center of the earth. This magnetic field is assumed by many geophysicists to arise from electric currents in the core, and they believe that the mechanism that generates these electric currents can be likened to a self-exciting dynamo driven by Earth's rotation. Although it is known that certain minerals such as magnetite and hematite (compounds of iron) are strongly magnetic, all rocks are magnetic to some degree because of the magnetic minerals that they contain. Sedimentary rocks may become magnetized when small magnetic mineral grains settle out during sedimentation, and igneous rocks may become magnetic from the formation of iron minerals during the cooling of lava or magma. As each particle becomes magnetized, and before solid rock has formed, these minute natural magnets line up like tiny compass needles, each pointing toward the earth's magnetic pole. Once the sediment or magma has hardened to form rock, millions of tiny "compass needles" continue to point toward Earth's magnetic field. Still more important, these minerals remain magnetized even after hundreds of millions of years, thus retaining a "memory" of the magnetic field as it was at the time and place the rocks were formed.

In younger rocks—say, those hardened from a recent lava flow—the magnetic mineral grains are aligned in the direction of the earth's magnetic field at that time. But magnetized material in older rocks may point in quite dif-

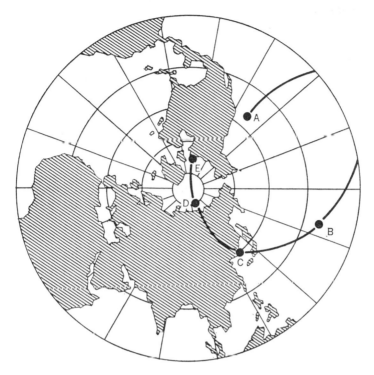

FIGURE 23 Reproduced by permission from *Investigating the Earth,* Earth Science Curriculum Project, Houghton Mifflin Company, 1967. Approximate position of the earth's North Magnetic Pole as determined from analysis of rocks from different ages. (A) Precambrian pole, (B) Cambrian pole, (C) Permian pole, (D) Cretaceous pole, and (E) present pole.

ferent directions from that of the present magnetic field. Consequently, it can be assumed that when the older rocks were formed, the earth's magnetic field was in a different direction than it is today. What do these fossil magnets tell us? First, they provide evidence that the earth's magnetic field has gradually changed position throughout geologic time. For example, during various chapters of Earth history, the North Magnetic Pole may have been in such unexpected locations as Korea, Arizona, and the North Pacific Ocean. Even more astonishing is evidence that suggests that the North and South Magnetic Poles have completely changed positions in the geologic past. Indeed, recent studies indicate that the earth's magnetic field has reversed at least nine times in the past 3,600,000 years!

What has this to do with drifting continents and spreading sea floors? Geophysicists have investigated the paleomagnetic properties of rocks of various ages and have constructed a polar wandering curve. They assume that if the continents have *not* changed positions, the polar curves for all continents would remain the same. But if the continents *have* drifted with respect to each other, the curves should be different. As interpreted by many geoscientists, paleomagnetic analyses clearly yield different polar curves for the different continents, and this suggests that the continents have indeed shifted.

Many opponents of continental drift have criticized the jigsaw puzzle fit of the Atlantic coastlines as coincidence or imagination. And not without reason, for there are quite a few missing pieces in this "continental jigsaw puzzle." But within recent years the "drifters" have been supported by a most objective and quantitative ally: the computer. The computer approach was first used in 1964 by Sir Edward C. Bullard, a geophysicist at Cambridge University. Although the Atlantic shorelines of South America and Africa do resemble complementary pieces of a fractured block, Bullard thought that the match might be even closer along the five-hundred-fathom line of the continental slope. This proved to be correct. Putting together North and South America, Africa, Greenland, and all the other islands, Bullard's computer study fitted the landmasses together with astonishing accuracy: less than one degree error on the average. A similar computer study was made in 1969 by two geological oceanographers of the Environmental Science Services Administration's Atlantic Oceanographic Laboratories. Using oceanographic data obtained from the U. S. Coast and Geodetic Survey Ship *Oceanographer,*

Walter Sproll and Robert S. Dietz have made what they call a "precise fit" between Australia and Antarctica. To do this they wrote a computer program using numbers and mathematical equations that indicated the geographic configurations and locations of the continental slope of Australia and Antarctica. After finding the closest fit position of the two continents, the computer then programmed a plotting machine to draw up a map of the fit. Their results showed that the total area of misfit between the two continents consisted of an area only about the size of Illinois. What is more, this amount of misfit can probably be reduced when more is known about the continental margin of Antarctica. However, the continental margins will never match perfectly, for the edges of these landmasses have undergone considerable geologic change since the proposed continent of Gondwanaland broke apart.

Physics and computer science are but two areas from which the drift theory has gained support. Biology, especially biogeography (the study of the geographic distribution of organisms), has also risen to the defense of Wegener's theory. As early as 1858, the similarity of fossil plants in America and Europe prompted Antonio Snider to suggest that all continents had once been part of a single landmass. Later discoveries have also proved that the plants and animals of the southern continents are remarkably similar in a number of respects. These areas are now separated by thousands of miles of deep ocean, which would normally have impeded the migration of land and fresh-water organisms. But if these continents had once been united, the distribution of these organisms is much easier to explain. The most commonly cited fossil evidence consists of the *Glossopteris* flora and a small, carnivorous reptile called *Mesosaurus*. The *Glossopteris*, or "tongue fern," is commonly found associated with widespread glacial deposits that blanket much of central and southern Africa, Australia, New Zealand, parts of South America, and peninsular India. Regardless of how these plants were dispersed, botanists agree that the "tongue ferns" could not possibly have spread across the broad expanse of open sea that now separates these areas. Thus, the *Glossopteris* flora—plus the widely distributed glacial material in which the plant fossils occur—clearly suggest a once-continuous landmass and further uphold the basic idea of continental drift. As for *Mesosaurus,* this meat-eating reptile is found in rocks of the same age on both sides of the Atlantic. Paleobiologists do not think it is possible that the forms found on the two continents could have evolved independently. Nor is it likely that this fresh-water creature could have survived a trip across the open ocean.

Additional fossil evidence to support continental drift has been found in the Antarctic region. Discovered in late 1967 by geologists from Ohio State University's Institute for Polar Studies, one of these fossils represents the jawbone of an ancient fresh-water amphibian called a labyrinthodont. Such fossils had previously been found in Australia and South Africa, and their presence in Antarctica can be considered as yet another link in the chain of evidence that indicates that the continents of the Southern Hemisphere were once united. Then—in late 1969—the remains of a reptile called *Lystrosaurus* were discovered in Antarctica. Fossils of this 225 million-year-old creature had previously been collected in Africa and South America, and its occurrence in Antarctica lends further credibility to the idea of drifting continents. Zoologists agree that these reptiles and fresh-water amphibians could not possibly have survived an intercontinental ocean trip, and it is inconceivable that they underwent separate evolution on each of the three continents.

More recently, biologists have added a new piece of evidence from the living world. While dissecting a fish from Lake Surprise in Tasmania, a biologist discovered a crustacean parasite, *Dolops tasmanianus.* Although there is nothing unusual about finding parasites in fish, this was an astonishing find, for *Dolops* has never been reported outside of South America and Africa. According to zoologists who have studied *Dolops,* the distribution of this animal must assume the existence of land connections between South America, Africa, Tasmania—and, quite conceivably —Australia.

Chemistry is a veritable cornerstone of geological investigation, and it has also added much to our knowledge of continental drift. Geochemists who have studied the chemical characteristics of rocks on the ocean floor and opposing continents have found significant similarities between them. By means of radiometric dating, they have also learned that rocks in such widely separated locations as India, Australia, South America, and Africa can be closely related to each other on the basis of their geologic age. The implication is, of course, that the matching rock sections in these widely separated areas were once conterminous bodies.

But within recent years, the rapidly blossoming science of oceanography has amassed the most compelling array of evidence to support continental drift. The contributions of geological oceanographers have helped to overcome the greatest single objection to the theory: What was the mechanism that made drifting possible? During the last decade there has been intensive exploration of the ocean basins, and much has been learned about their origin, composition, and structure. Among the more important findings is much evidence to suggest that the ocean floors are not stationary, but are slowly expanding. Careful and systematic mapping indicates that this spreading originates at the Mid-Atlantic Ridge, part of a globe-girdling, forty-thousand-mile-long chain of submarine mountains that twists and branches through the Pacific, Indian, and Atlantic Ocean basins. This oceanic ridge, which is as much as five miles high and several hundred miles wide, is composed of rugged submarine lava flows capped by a relatively thin layer of oceanic sediments. Steplike slopes are found on each side of the Mid-Atlantic Ridge and a deep, narrow trench, or rift, in the top of the ridges. This seems reasonable, for the mid-oceanic rifts are bordered by island arcs with many active volcanoes, and are regions of much earthquake activity.

The molten rock that has boiled up from the earth's interior has apparently poured out through the central rift in the ridge. And, with each outpouring of lava, the solid rock on either side of it has been pushed progressively farther away. Although the mid-ridge trench is clearly a zone of tension and the sea floor is tearing apart along an ever-widening crack, the rift does not extend downward indefinitely. Rather, this fissure is continually being filled by lava welling into it from below. The newly made rock formed from this lava is found in and near the trench, with progressively older parallel bands on either side of it. This system of increasingly older parallel bands of rock has led to the assumption that the ocean basin is continually widening and being renewed by molten rock flowing from the central trench.

How do we know that the rocks are progressively older away from the ridge? This has been substantiated by geophysical surveys of the fossil magnetism in the bands of rock on each side of the oceanic ridge. Geophysicists reasoned that if sea-floor spreading actually has occurred, lava extruded on the ocean floor adjacent to the ridge during the past seven hundred thousand years would have normal polarity, for Earth's magnetic field has remained at the same place during this period of time. By the same token, the lava flows of seven hundred thousand to 850,000 years ago would lie in a parallel band next to the first band and its paleomagnetism would be reversed, for this was a time of reversed polarity. To test this theory, the seagoing geophysicists towed a magnetometer (an instrument that measures magnetic intensity) across a ridge in the Indian Ocean. True to their expectations, they recorded alternating bands of normal and reversed magnetic polarity as they moved farther away from the ridge. The magnetometer readings showed that the magnetic field was alternately high, then low, because the rocks over which the magnetometer was towed, having either normal or reversed polarity, added to or subtracted from Earth's main magnetic field. These studies, which were also made of other oceanic ridges, indicate that the ocean bottom on either side of the ridge was magnetized in a remarkably symmetrical stripelike pattern. In one sense, these linear bands of alternating normal and reversed polarity have recorded the development of the ocean floor much as the rings in a stump indicate the pattern of growth in a tree. Recent studies of magnetic reversals, together with radiometric age determinations of the oceanic rocks, indicate that there have been at least 171 such magnetic reversals in the last seventy-six million years—a relatively short time, geologically speaking. Even more important is the fact that these magnetic anomalies reveal that the ocean floor is spreading from one to six centimeters per year.

Many geoscientists believe that sea-floor spreading is the work of thermal convection currents within the earth's mantle (FIGURE 15). It is known that convection currents are generated when gases and liquids are heated, for when heat is applied, the cooler portion of a liquid sinks because it is more dense than the heated portion. This pushes up the heated part, and as the rising liquid reaches the surface it loses its heat, becomes more dense, and then moves down again. This continual exchange of heat results in circulating currents of convection cells that churn through the liquid as long as heat is applied. It is generally agreed that the heat responsible for convection cells in the mantle is caused by the steady disintegration of radioactive elements such as uranium. This nuclear heat is contained in the mantle because the overlying rocks conduct heat very slowly and act as an insulating "blanket." This permits the buildup of heat that causes the mantle to expand and convect upward. In time, some of this molten rock material will reach the surface by means of rifts in the mid-ocean ridges. As the lava pours out on the ocean floor, it moves the earlier formed rock along the crust and away from the ridge.

The concept of sea-floor spreading suggests that the

Atlantic Ocean basin is becoming wider as the continents are being pushed away from the Mid-Atlantic Ridge. This means, in turn, that the Pacific Ocean basin is presumably becoming smaller. The excess material in the shrinking Pacific basin is believed to be forced downward through a series of oceanic trenches, where it is pulled into the mantle by the descending segments of adjacent convection cells.

What does sea-floor spreading have to do with continental drift? A number of earth scientists believe that those portions of the convection cells that move parallel to the surface have caused the surface of the sea bottom to move. By combining the mechanism of drift with that of convection in the mantle, it is assumed that mid-ocean ridges are located above the junctions of upwelling currents that tend to spread the oceanic crust apart. This has produced the broad arch of the ridge, the rift along its crest, and the volcanic and earthquake activity associated with it. It is further assumed that convection currents rise along each flank of an ocean ridge; then, moving laterally beneath the ocean bottom, they eventually sink at the nearest continental margin, dragging part of the ocean floor down into the mantle. The currents that sink at the margins of continents are believed to be responsible for the presence of the deep ocean trenches and the island arcs. This is especially applicable to the Pacific Ocean, which is bordered by trenches, but less so for those oceans that lack marginal trenches. The continents are also assumed to be located over the approximate juncture of descending currents. Thus, according to this theory the ocean floor acts as a great "conveyor belt" that has carried the continents along with it. The magnetic lineations on the ocean floor support this movement and might be likened to "footprints" left by the continents as they moved to their present positions.

Geological data also support the geophysical evidence for spreading sea floors. During 1968 and 1969, the oceanographic research vessel *Glomar Challenger* took a series of long rock cores from the bottom of the Atlantic and Pacific Oceans. Part of the Deep Sea Drilling Project, the purpose of the sea-floor sampling was to determine the age of the ocean basins and the history of their formation and development. Marine geologists assumed that if sea-floor spreading has occurred and new oceanic crust really is being created in the central part of the Atlantic Ocean, then it should follow that the oldest ocean crust with the oldest sediments overlying it should lie adjacent to continental margins on both sides of the Atlantic. Conversely, progressively younger crust with equally young sediments overlying it should be found toward the central part of the ocean. Oceanographers felt that this hypothesis could be tested by drilling a number of holes across the ocean basin and obtaining samples of the sediment that tops the oceanic crust. The age of the sediments could then be determined by the presence of fossilized micro-organisms that died and were entombed in the sediments. The results? *Glomar Challenger* sank test holes at nine sites in the North and South Atlantic to sample the sediments overlying the crustal material. Without exception, the results indicated that new ocean crust was being emplaced at the crest of the Mid-Atlantic Ridge. As expected, the oldest microfossils were recovered in sediments near the continental margin. Although the core barrel did not quite penetrate the base of the sedimentary layer, geophysical data support the contention that the Atlantic Ocean basin is a relatively young geologic feature. Its age has been set at about two hundred million years, a figure that is quite compatible with other scientific estimates as to when the continents may have separated. The rate of sea-floor spreading was also calculated by comparing the age of the oldest ocean sediments and the distance from the crest of the ridge. Again, their estimate of two to three centimeters per year is in agreement with spreading rates predicted on the basis of other geological and geophysical studies. Assuming that geophysical data for other parts of the oceans are also correct, then rates as high as six centimeters from the ridge crest are possible. This means, then, that two points on either side of the ridge could be separated a total of twelve centimeters per year. Although this amount of movement may seem insignificant, it is quite rapid within the framework of geologic time. Consider the fact that during our lifetime the ocean floor may have spread and the continents moved a distance roughly proportionate to our height. Consider also the fact that we have lived but a relatively short time, while the oldest known continental rocks are about three billion years of age. In other words, there has been more than enough time for the continents to have reached their present positions.

Structural geologists are interested in continental drift as a possible explanation for the origin of folded mountain ranges. Wegener and his supporters thought that the world's mountain belts were formed as the leading edge of the great continental "rafts" became buckled because of resistance from the underlying plastic "sea" in which they were drifting. Current thinking does not call for continental "rafting" alone. Instead, many earth scientists believe that the upper part of the upper mantle, the ocean floor, and the continents are all drifting. This does not prove, of course, that the continents have drifted. But the evidence that has been gathered in the past five to ten years certainly appears to favor the concept of drifting continents and spreading sea floors.

Curiously enough, the final proof of continental drift may come from outer space. Among the instruments left on the moon by Apollo 11 astronauts was a reflecting prism. This has been the target of a laser beam that astronomers have used to determine the exact distance from the earth to the moon. Future measurements will be used to detect any increase in the earth-moon distance that could be indicative of lateral continental movements. Similar evidence for continental separation may also come from satellite geodesy. The Smithsonian Astrophysical Observatory has established a worldwide network for the optical tracking of satellites. Using laser rather than optical measuring techniques, scientists should be able to pinpoint sites on the earth's surface to within an error of about ten centimeters. If, as has been suggested, the continents are moving apart at a rate of one to six centimeters per year, such movement could be documented in a rather reasonable length of time. Thus, after a period of some 350 years, geoscientists may be approaching the solution of one of Earth's greatest mysteries. To do this scientists have used virtually every tool in their "bag of tricks" and have applied the basic concepts and research techniques of physics, chemistry, biology, astronomy, and mathematics. Now —thanks to investigations that range from Bacon's "jigsaw puzzle" approach of the early seventeenth century to the use of the most sophisticated instruments of the Space Age —this long-standing controversy may soon be settled once and for all.

9

GEOLOGY AND THE FUTURE

The manner in which earth scientists have attacked the problem of continental drift is typical of what some of them are calling the "new" geology. Not that the problems are new—many, such as continental drift and mountain building, have been puzzled over for centuries. There is, instead, a new thrust in the geological sciences, and at the cutting edge of this research are new techniques and tools that were undreamed of a few years ago. This geoscience "revolution" is not coming too soon. With increasing demands for mineral resources, geologists have been called upon to greatly intensify their search for these valuable products. Thanks to recent developments in science and technology, it is now possible to find and recover mineral resources that were unattainable a decade or two ago. But increased availability has brought about increased demand as engineers and scientists have put these earth materials to new and unforeseen uses.

Thus, despite the great strides that have been made in geologic exploration, we are barely meeting the continually expanding requirements of modern technology. The ever-increasing need for mineral resources, plus the realization that older and more conventional methods are not yielding the results that they once did, has been responsible for many of these exciting new advances in geologic research.

Will the "new" geology permit us to develop the geologic "know-how" that will be needed to replenish our rapidly dwindling natural resources? The answer must clearly be "Yes." Only a better understanding of the earth's crust will make it possible for man to develop and expand in this area in the years ahead.

And so today's earth scientists are taking a new—and much closer—look at this ancient planet. This new breed of geoscientist is employing an ever-increasing array of geologic tools that can be used on land, sea, and in the air; in addition, photographs taken from space vehicles are providing unparalleled views of the geology and geography of many parts of the world. These remarkable pictures— any one of which might cover an area of thousands of square miles—are the bases for expanded studies of the culture, resources, surface features, and geology of the areas concerned. So promising is geologic exploration from space that the United States Geological Survey and the National Aeronautics and Space Administration are planning to launch their own observational satellites. Equipped with remote sensing instruments such as ultraviolet and infrared sensors, radar scanning systems, as well as wide-angle television and still cameras, these self-contained, orbiting laboratories will quickly and accurately provide information that is otherwise unobtainable.

Practitioners of the "new" geology are especially interested in the oceans. These seagoing geologists—like all geologists—are curious. And the mere fact that almost 71 percent of the earth's crust is obscured by water makes them all the more eager to learn more about the ocean floor. But they also know that their research may be of practical value, for more than five billion dollars' worth of oil, natural gas, and minerals have been extracted from rocks of the rather narrow continental shelves that border the United States. And yet we have hardly scratched and are barely familiar with the great mineral wealth that is believed to be located on the floor of the deeper ocean basins.

Other geologists—or planetologists—are studying our sister planets, paving the way for their future exploration and attempting to learn more about their relation to Earth. Astronauts have visited the moon, and each time returned with a precious cargo of lunar rocks. Selenologists (specialists in lunar geology) have studied these in order to learn more about the relation of Earth to the moon and to determine their age and origin. Much attention is also being directed to learning more about the powerful forces at work within the earth's interior. Deep test holes are being drilled, lava samples removed from volcanic craters, and earthquake movements are being carefully monitored. The results of these and similar investigations might pave the way for the prediction of such natural phenomena as earthquakes and volcanic eruptions.

Yet even as geoscientists probe the earth's interior, solve the secrets of the ocean deeps, and scan Earth's surface from the reaches of outer space, they do not ignore the rocks beneath their feet. "Field boot" geology is still a "must," and geologists rely—as they have for two hundred years—on the pick and hammer, hand lens, maps, and compass. Although satellites may be used to compile maps, and remote sensing devices may reveal the nature of earth's surface materials, their findings must eventually be checked by the field geologist. Samples must be collected and documented and later analyzed in the laboratory. Depending upon the information desired, geological samples may be tested in many different ways. Many of the instruments used—for example, the electron microscope, x-ray machine, and mass spectrometer—were originally developed for other basic sciences, but have been especially adapted for geologic investigation.

But far more important, the "new" geology is addressing itself to the problem of man and his physical environment. Man is a creature of the earth, and Earth is responsible for him. But the environmental geologist realizes that man is also responsible for the earth. This old planet has provided a suitable environment for the evolution of the human race and made available the necessary energy sources and raw materials for civilization. Earth has been good to man, and he is the "caretaker" of our planetary home. How have we cared for the earth and the natural environment that makes our existence possible? We have polluted its atmosphere, contaminated its water, mismanaged its mineral resources, and overfarmed its soils. Worse yet, man continues to increase in unprecedented numbers, and this steady increase in population gives rise to still more environmental problems. The environmental geologist is dedicated to the conservation of natural resources and the application of geology to human needs, for he is painfully aware that Earth does not have an unlimited supply of potable water, raw mineral resources, and usable land area.

Because many of the problems confronting environmental geologists are particularly acute in the cities, special attention is being given to studies in urban geology. The disposal of wastes and air and water pollution are prime targets, of course. But equally important is the insurance of proper construction practices consistent with local geologic conditions. Consider, for example, the disastrous landslides that are plaguing parts of California. If residential housing contractors and city planners had given more thought to slope stability and the nature of local bedrock, much of this destruction could have been avoided. Other great losses have been sustained because of poor engineering and construction of dams. A tragic example is the collapse of the Mapasset Dam in southern France. If one side of the dam had not been unknowingly anchored in an unstable, clay-filled fracture, 344 lives would probably have been spared. Closer to home in 1963 was the failure of the Baldwin Hills Reservoir in Los Angeles. Astonishingly enough, this huge water reservoir was built in an area known to be underlain by an active fault. Movements along this fracture eventually weakened the dam to the extent that water leaking through cracks led to the sudden collapse of the dam. Although property loss was great, there was enough advance warning to prevent a large loss of life. Needless to say, geologic advice is especially needed in areas that are normally prone to landslides and earthquake activity.

But the interest of the environmental geologist is not confined to recognizing geologic hazards and preventing natural disasters. He is also interested in reducing marine erosion in coastal areas, protecting valuable soils, and making the best use of our land and mineral resources. More recently geologists have united with medical workers to determine the relation of the chemistry of rocks, soils, plants, and water to health and disease. Geomedical research indicates that the occurrence of certain chronic diseases and health problems shows definite geographic patterns throughout the world. For example, the distribution of cancer deaths was plotted on a geologic map of West Devonshire, England, and the results clearly indicated a high concentration of cancer victims among persons living on certain geologic formations. Subsequent chemical analyses revealed that the areas with high death rate from cancer had unusually high concentrations of lead in the soil. This led to the discovery that vegetables grown in these soils also had a high lead content, and the consumption of these vegetables would have transmitted the lead to those who ate them. The absence of certain elements can also be detrimental to residents living in certain geologic provinces. A deficiency in zinc may slow down the maturity of the human body, and a sufficient amount of copper is needed to promote the proper development of connective tissue in man. Likewise, a lack of chromium produces an effect that might stimulate diabetes. Although it is true that the causes are not clearly understood, there is much evidence to suggest that some health problems are at least partially related to environmental geochemistry. As research in medical geology increases, other relations between disease and geologic environment will doubtless be discovered.

With each passing year increasingly heavier demands

will be made on our natural mineral resources, and our environment will be further strained. In short, Earth's invitation to the ancient caveman has become Earth's challenge to modern man. Can *Homo sapiens* meet the challenge? Or will he—like the dinosaurs, dodos, and countless other species—be recorded as just another extinction in Earth's rocky ledger? No one knows, of course, for time alone will tell. But one thing seems certain: To ignore Earth's challenge will be fatal, for it can only lead to irreparable damage of our already plundered planet. The burden of this confrontation weighs heavily on and is especially pertinent to the geological sciences. If, as James Hutton so aptly suggested, "The present is the key to the past," geology will not only meet the challenge—it may well emerge as the most critical science of our times.

SELECTED READINGS

Hopefully, this *Invitation to Geology* has whetted the reader's appetite for additional information about planet Earth. The following list includes selected references of many types—any one of which will provide further reading on the various phases of geology. This list is by no means all-inclusive, and many other interesting and informative publications may be found in public, school, and college libraries. Wherever possible, the paperback edition of a book is listed and designated by an asterisk. Those not so marked are available only in hardcover.

TEXTBOOKS

General Geology

Bates, Robert L., and Sweet, Walter C., *Geology: An Introduction.* D. C. Heath & Company, Boston (1966).

Eardley, Armand J., *General College Geology.* Harper & Row, New York (1965).

Fagan, John J., *View of the Earth: An Introduction to Geology.* Holt, Rinehart & Winston, New York (1965).

Putnam, William C., *Geology.* Oxford University Press, New York (1964).

Shelton, John S., *Geology Illustrated.* W. H. Freeman and Co., San Francisco (1966).

Stokes, William L., and Judson, Sheldon, *Introduction to Geology: Physical and Historical.* Prentice-Hall, Englewood Cliffs, New Jersey (1968).

Zumberge, James H., *Elements of Geology* (2nd ed.). John Wiley & Sons, New York (1963).

Physical Geology

Emmons, William H., Allison, Ira S., Stauffer, Clinton R., and Thiel, George A., *Geology: Principles and Processes* (5th ed.). McGraw-Hill Book Company, New York (1960).

Gilluly, James, Waters, Aaron C., and Woodford, A. O., *Principles of Geology* (3rd ed.). W. H. Freeman and Co., San Francisco (1968).

Holmes, Arthur, *Principles of Physical Geology* (2nd ed.). The Ronald Press Company, New York (1965).

Holmes, Doris L., *Elements of Physical Geology*. The Ronald Press Company, New York (1969).

Leet, L. Don, and Judson, Sheldon, *Physical Geology*. Prentice-Hall, Englewood Cliffs, New Jersey (1965).

Longwell, Chester R., Flint, Richard Foster, and Sanders, John E., *Physical Geology*. John Wiley & Sons, New York (1969).

Rogers, John J. W., and Adams, John A. S., *Fundamentals of Geology*. Harper & Row, New York (1966).

Spencer, Edgar W., *Basic Concepts of Physical Geology*. Thomas Y. Crowell, New York (1962).

Historical Geology

Clark, Thomas H., and Stearn, Colin W., *Geological Evolution of North America*. The Ronald Press Company, New York (1968).

Dunbar, Carl O., and Waage, Karl M., *Historical Geology* (3rd ed.). John Wiley & Sons, New York (1969).

Kay, Marshall, and Colbert, Edwin H., *Stratigraphy and Life History*. John Wiley & Sons, New York (1965).

Kummel, Bernhard, *History of the Earth: An Introduction to Historical Geology*. W. H. Freeman and Co., San Francisco (1961).

Moore, Raymond C., *Introduction to Historical Geology*. McGraw-Hill Book Company, New York (1958).

Spencer, Edgar W., *Basic Concepts of Historical Geology*. Thomas Y. Crowell, New York (1962).

Stokes, William L., *Essentials of Earth History: An Introduction to Historical Geology*. Prentice-Hall, Englewood Cliffs, New Jersey (1966).

Woodford, A. D., *Historical Geology*. W. H. Freeman and Co., San Francisco (1965).

INTRODUCTORY READINGS AND "OUTLINES"

* Cox, Dennis P. and Helen R., *Introductory Geology: A Programmed Text.* W. H. Freeman and Co., San Francisco (1965).

* Foster, Robert J., *Geology.* Charles E. Merrill Books, Columbus, Ohio (1966).

* Leet, L. Don and Florence J., *The World of Geology.* McGraw-Hill Book Company, New York (1961).

* Matthews, William H., III, *Geology Made Simple.* Doubleday & Company, Garden City, New York (1967).

* Pearl, Richard M., *Geology: An Introduction to Principles of Physical and Historical Geology.* Barnes & Noble, New York (1960).

* ———, *Geology Simplified.* Barnes & Noble, New York (1967).

Scientific American. Resource Library. *Readings in the Earth Sciences.* W. H. Freeman and Co., San Francisco (1969).

* White, John F. (Ed.), *Study of the Earth: Readings in the Geological Sciences.* Prentice-Hall, Englewood Cliffs, New Jersey (1962).

General (Nontechnical)

Beiser, Arthur, and the Editors of *Life, The Earth.* Time-Life Books, New York (1962).

Clayton, Keith, *The Crust of the Earth: The Story of Geology.* The Natural History Press, Garden City, New York (1967).

Dunbar, Carl O., *The Earth.* World, New York (1966).

* Gamow, George, *A Planet Called Earth.* The Viking Press, New York (1963).

Mather, Kirtley F., *The Earth Beneath Us.* Random House, New York (1964).

Matthews, William H., III, *A Guide to the National Parks: Their Landscape and Geology* (Vol. I, *The Western Parks;* Vol. II,· *The Eastern Parks*). The Natural History Press, Garden City, New York (1968).

* Page, Lou W., *The Earth and Its Story.* American Education Publications, Columbus, Ohio (1961).

* Rapport, Samuel, and Wright, Helen (Eds.), *The Crust of the Earth.* The New American Library, New York (1962).

Shimer, John A., *This Changing Earth: An Introduction to Geology.* Harper & Row, New York (1967).

* Viorst, Judith, *The Changing Earth.* Bantam Books, New York (1967).

* Wyckoff, Jerome, *Geology: Our Changing Earth Through the Ages.* Golden Press, New York (1967).

———, *Rock, Time, and Landforms.* Harper & Row, New York (1966).

EARTH MATERIALS AND PROCESSES

General

* Cailleux, André, *Anatomy of the Earth.* McGraw-Hill Book Company, New York (1968).

Fraser, Ronald G. J., *The Habitable Earth.* Basic Books, New York (1965).

* Spar, Jerome, *Earth, Sea, and Air: A Survey of The Geophysical Sciences.* Addison-Wesley, Palo Alto, California (1962).

Earth Materials

Fenton, Carroll L. and Mildred A., *The Rock Book.* Doubleday, Garden City, New York (1940).

Gallant, Roy A., and Schuberth, Christopher J., *Discovering Rocks and Minerals.* The Natural History Press, New York (1967).

Hurlbut, Cornelius S., Jr., *Minerals and Man.* Random House, New York (1968).

* Hurlbut, Cornelius S., and Wenden, Henry E., *The Changing Science of Mineralogy.* D. C. Heath & Company, Boston (1967).

* Pearl, Richard M., *How to Know the Rocks and Minerals.* Signet, New York (1957).

———, *Rocks and Minerals.* Barnes & Noble, New York (1956).

Pough, Frederick H., *A Field Guide to the Rocks and Minerals* (3rd ed.). Houghton Mifflin, Boston (1960).

Sinkankas, John, *Mineralogy for Amateurs.* D. Van Nostrand, Princeton, New Jersey (1964).

* Skinner, Brian J., *Earth Resources*. Prentice-Hall, Englewood Cliffs, New Jersey (1969).

Geologic Processes

* Adams, William M., *Earthquakes: An Introduction to Observational Seismology*. D. C. Heath & Company, Boston (1964).

Bullard, Fred M., *Volcanoes: In History, in Theory, in Eruption*. University of Texas Press, Austin, Texas (1962).

Engle, Eloise, *Earthquake! The Story of Alaska's Good Friday Disaster*. John Day, New York (1966).

Herbert, Don, and Bardossi, Fulvio, *Kilauea: Case History of a Volcano*. Harper & Row, New York (1968).

Heintze, Carl, *The Circle of Fire: The Great Chain of Volcanoes and Earth Faults*. Meredith Press, New York (1968).

* Hodgson, John H., *Earthquakes and Earth Structure*. Prentice-Hall, Englewood Cliffs, New Jersey (1964).

Iacopi, Robert, *Earthquake County: How, Why, and Where Earthquakes Strike California*. Lane Magazine & Book Co., Menlo Park, California (1964).

* Leet, L. Don and Florence, *Earthquake: Discoveries in Seismology*. Dell Publishing Co., New York (1964).

Roberts, Elliott, *Our Quaking Earth*. Little, Brown and Company, Boston (1963).

* Summer, John S., *Geophysics, Geologic Structures and Tectonics*. William C. Brown Company, Dubuque, Iowa (1969).

Vaughan-Jackson, Genevieve, *Mountains of Fire: An Introduction to the Science of Volcanoes*. Hastings House, New York (1962).

Wilcoxson, Kent H., *Chains of Fire: The Story of Volcanoes*. Chilton Books, Philadelphia (1966).

Landforms and the Earth's Surface

* Bloom, Arthur L., *The Surface of the Earth*. Prentice-Hall, Englewood Cliffs, New Jersey (1969).

* Dury, G. H., *The Face of the Earth*. Penguin Books, Baltimore, Maryland (1966).

Farb, Peter, *Face of North America*. Harper & Row, New York (1962).

Shimer, John A., *This Sculptured Earth, The Landscape of America.* Columbia University Press, New York (1959).

EARTH HISTORY AND FOSSILS

* Barnett, Lincoln and the Editors of *Life, The World We Live In.* Golden Press, New York (1956).

Beerbower, James R., *Search for the Past: An Introduction to Paleontology* (2nd ed.). Prentice-Hall, Englewood Cliffs, New Jersey (1968).

Carrington, Richard, *The Story of Our Earth.* Harper & Row, New York (1956).

* Clark, David L., *Fossils, Paleontology and Evolution.* William C. Brown Company, Dubuque, Iowa (1968).

Colbert, Edwin H., *The Age of Reptiles.* W. W. Norton & Company, New York (1965).

———, *Dinosaurs, Their Discovery and Their World.* Dutton, New York (1961).

* ———, *Evolution of the Vertebrates: A History of Backboned Animals.* John Wiley & Sons, New York (1955).

Ericson, David B., and Goesta, Wollin, *The Deep and the Past.* Alfred A. Knopf, New York (1964).

Fenton, Carroll L. and Mildred A., *The Fossil Book.* Doubleday, Garden City, New York (1965).

* Hotton, Nicholas, III, *Dinosaurs.* Pyramid Publications, New York (1963).

* ———, *The Evidence of Evolution.* American Heritage, New York (1968).

* Laporte, Leo F., *Ancient Environments.* Prentice-Hall, Englewood Cliffs, New Jersey (1968).

* Matthews, William H., III, *Fossils: An Introduction to Prehistoric Life.* Barnes & Noble, New York (1962).

* McAlester, A. Lee, *The History of Life.* Prentice-Hall, Englewood Cliffs, New Jersey (1968).

Moore, Ruth, *Man, Time, and Fossils.* Alfred A. Knopf, New York (1965).

Simak, Clifford D., *Trilobite, Dinosaur and Man: The Earth's Story.* St. Martin's Press, New York (1966).

* Simpson, George G., *Life of the Past: An Introduction to Paleontology.* Yale University Press, New Haven, Connecticut (1953).

* Stirton, R. A., *Time, Life, and Man: The Fossil Record.* John Wiley & Sons, New York (1963).

GEOLOGIC TIME

Berry, William B. N., *Growth of a Prehistoric Time Scale.* W. H. Freeman and Co., San Francisco (1968).

* Eicher, D. L., *Geologic Time.* Prentice-Hall, Englewood Cliffs, New Jersey (1968).

* Faul, Henry, *Ages of Rocks, Planets and Stars.* McGraw-Hill Book Company, New York (1968).

* ———, *Nuclear Clocks.* U. S. Atomic Energy Commission, Oak Ridge, Tennessee (1967).

* Harbaugh, John W., *Stratigraphy and Geologic Time.* William C. Brown Company, Dubuque, Iowa (1968).

* Hurley, Patrick M., *How Old is the Earth?* Doubleday, Garden City, New York (1959).

GEOLOGY AND THE SEA

* Carson, Rachel L., *The Sea Around Us.* The New American Library, New York (1961).

* Cowen, Robert C., *Frontiers of the Sea.* Bantam Books, New York (1960).

* Gross, M. Grant, *Oceanography.* Charles E. Merrill Books, Columbus, Ohio (1967).

Keen, M. J., *An Introduction to Marine Geology.* Pergamon Press, New York (1966).

* Long, John E., *New Worlds of Oceanography.* Pyramid Publications, New York (1965).

* Menard, Henry W., *Anatomy of An Expedition.* McGraw-Hill Book Company, New York (1969).

* *Scientific American, The Ocean.* W. H. Freeman and Co., San Francisco (1969).

Shepherd, Francis P., *The Earth Beneath the Sea* (rev. ed.). The Johns Hopkins Press, Baltimore, Maryland (1967).

Shepherd, Francis P., and Dill, Robert F., *Submarine Canyons and Other Sea Valleys.* Rand McNally & Co., Chicago, Illinois (1966).

* Yasso, Warren E., *Oceanography: A Study of Inner Space.* Holt, Rinehart & Winston, New York (1965).

GEOLOGY AND MAN

Flawn, Peter T., *Environmental Geology,* Harper & Row, New York (1970).

Matthews, William H., III, *Science Probes the Earth: New Frontiers of Geology.* Sterling Publishing Co., New York (1969).

* Roy, Chalmer J., *The Sphere of the Geological Scientist.* American Geological Institute, Washington, D.C. (1963).

* Turner, Daniel S., *Applied Earth Science.* William C. Brown Company, Dubuque, Iowa (1969).

HISTORY OF GEOLOGY

* Adams, Frank D., *The Birth and Development of the Geological Sciences.* Dover Publications, New York (1954).

Fenton, Carroll L. and Mildred A., *Giants of Geology.* Doubleday, Garden City, New York (1962).

* Geikie, Sir Archibald, *The Founders of Geology* (2nd ed.). Dover Publications, New York (1962).

Mather, Kirtley F., and Mason, Shirley L., *A Source Book in Geology.* McGraw-Hill Book Company, New York (1939).

Moore, Ruth, *The Earth We Live On: The Story of Geological Discovery.* Alfred A. Knopf, New York (1961).

THE EARTH IN SPACE

Asimov, Isaac, *The Double Planet.* Abelard-Schuman, New York (1960).

Branley, Franklin M., *The Earth: Planet Number Three.* Thomas Y. Crowell Company, New York (1966).

* Nicholson, Thomas D. (Ed.), *Astronomy Highlights.* The Natural History Press, Garden City, New York (1964).

Odishaw, Hugh, *The Earth in Space.* Basic Books, New York (1967).

* Stumpff, Karl, *Planet Earth.* The University of Michigan Press, Ann Arbor, Michigan (1959).

MISCELLANY

* Albritton, Claude C., Jr. (Ed.), *Fabric of Geology*. Freeman, Cooper & Co., San Francisco (1963).

* American Geological Institute, *Dictionary of Geological Terms*. Dolphin, New York (1962).

Cortwright, Edward M., *Exploring Space With a Camera*. NASA SP-168, U. S. Government Printing Office, Washington, D.C. (1968).

Fielder, Gilbert, *Lunar Geology*. Lutterworth Press, London, England (1965).

Larousse Encyclopedia of the Earth. Prometheus Press, New York (1961).

* Takeuchi, H., Uyeda, S., and Kanamori, H., *Debate About the Earth: Approach to Geophysics Through Analysis of Continental Drift*. Freeman, Cooper & Co., San Francisco (1967).

* Wegener, Alfred, *The Origin of Continents and Oceans*. Dover Publications, New York (1966).

INDEX